In the Land of Orpheus

In the Land
of Orpheus

Rural Livelihoods and
Nature Conservation in
Postsocialist Bulgaria

Barbara A. Cellarius

THE UNIVERSITY OF WISCONSIN PRESS

The University of Wisconsin Press
1930 Monroe Street
Madison, Wisconsin 53711

www.wisc.edu/wisconsinpress/

3 Henrietta Street
London WC2E 8LU, England

Library of Congress Cataloging-in-Publication Data

Cellarius, Barbara A.
    In the land of Orpheus : rural livelihoods and nature conservation in postsocialist
Bulgaria / Barbara A. Cellarius.
        p. cm.
    Includes bibliographical references and index.
    ISBN 0-299-20150-3 (hardcover : alk. paper)—ISBN 0-299-20154-6 (pbk. : alk. paper)
        1. Human ecology—Rhodope Mountains Region. 2. Nature conservation—Rhodope
Mountains Region. 3. Agriculture—Environmental aspects—Rhodope Mountains
Region. 4. Rhodope Mountains Region—Rural conditions. 5. Agriculture and state—
Bulgaria. 6. Environmental policy—Bulgaria. I. Title.
    GF642.B82R463 2004
    333.72'09499'7—dc22   2004007742

Contents

Illustrations

Figures

Photos

Tables

Acknowledgments

This book would not have been possible without the assistance and support of many organization, institutions, and individuals. My fieldwork in and about Bulgaria has been supported in part by a National Science Foundation (NSF) Ethnographic Research Training Grant, administered by the Department of Anthropology at the University of Kentucky; an NSF Dissertation Improvement Grant (SBR#-9629832); a grant from the Nonprofit Sector Research Fund of the Aspen Institute; and a grant from the International Research and Exchanges Board (IREX) with funds provided by the National Endowment for the Humanities and the U.S. Department of State, which administers the Title VIII Program. I was also funded by IREX to attend their Bulgarian Studies Seminar. Initial analysis and write-up was supported in part with a dissertation fellowship from the East European Committee of the American Council of Learned Societies and a tuition waiver from the University of Kentucky Graduate School. This manuscript was prepared with the generous intellectual and financial support of a postdoctoral fellowship at the Max Planck Institute for Social Anthropology, Halle/Saale, Germany. None of these organizations is responsible for the views expressed here.

While these institutions provided critical financial support, numerous people also helped out along the way. My institutional home-away-from-home in Bulgaria is the Institute of Sociology at the Bulgarian Academy of Sciences, Department of Communities and Social Stratification. At the Institute I particularly benefited from the friendship and assistance of Veska Kozhuharova-Zhivkova and Stanka Dobreva as they shared with me their knowledge of Bulgarian villages and helped me to negotiate the Bulgarian bureaucracy. I passed many pleasant hours and meals in conversation about anthropology and Bulgaria at the home of Professor Asen Balikci in Sofia, and through him I met several other scholars of the region. I could similarly drop in on several people at the Institute of Geography for friendly conversation, the answers to questions, and a cup of tea. I also benefited from discussions with and the companionship of several non-Bulgarian scholars of things Bulgarian, including Bob Begg, Chad Staddon, Darren Spreeuw, Gerald Creed, John Pickles, Mieke Meurs, Moyuru Matsumae, Nadine Zanow, Tim Pilbrow, Ulf Brunnbauer, and Virginia Howard. My thoughts about Bulgarian environmental NGOs have benefited from extended discussions with Chad Staddon. I was assisted in my research by three very capable young Bulgarians who served as research assistants and translators: Vichra Baeva, Bobbie Popova, and Kristina Donkova. Three teachers from the school in Zaburdo administered a household survey and helped collect data on land inheritance. Svetla Anachkova and Mona Dzhoubrova also assisted with translation and data collection. Although I consider all of these people friends, Bobbie and her family deserve special thanks for their help from the minute I first set foot in Bulgaria.

My landlords and their families in Zaburdo, Sofia, and Chepelare took me into their homes and went out of their ways to help me negotiate life in Bulgaria and to make me feel at home. Village officials including the school director, librarian, and mayor were also of assistance. I will not try to name all the Bulgarians who befriended me during my stay in their country, but thanks are due to all of them for making my stay in their country more pleasant and sometimes less traumatic as well. Particular thanks are due to the residents of the Rhodope communities I visited over the course of my research and to the numerous members and officials of nongovernmental organizations with whom I spoke in Bulgaria and elsewhere. Without them this book would not have been possible.

Turning to people in the United States, having an archaeologist for an uncle made anthropology seem a perfectly normal thing to study in college. The late Richard Jordan introduced me to ecological approaches in anthropology at Bryn Mawr College. Several faculty members in the Masters Program in Environmental Studies at The Evergreen State College encouraged me to pursue a doctorate, including John Perkins, Ralph Murphy, Tom Womeldorff, and Lin Nelson. The members of my advisory committee at the University of Kentucky—Peter Little, Endre Nyerges, John van Willigen, John Pickles, and Tom Greider—all read and commented upon earlier versions of this manuscript, as did Asen Balikci, Daniel Bates, Chris Hann, Yulian Konstantinov, and Mieke Meurs. Steve Reyna encouraged me to employ the Orpheus theme. One map was prepared by Dick Gilbreath of the University of Kentucky Cartography Lab. Part of chapter 4 is a revised version of material previously published in *Ethnology* in 2000 and reprinted here with permission of the journal's editors. And last but not least, my parents provided critical logistical support during my travels around the globe.

Many thanks to everyone for their help and support! Of course, I take sole responsibility for the views expressed here and any errors that have crept into the text.

In the Land of Orpheus

Introduction

According to ancient Greek mythology, the Thracian musician Orpheus forsook human company after losing his young bride to the underworld and set out with his lyre to wander the pastures and forests of the Rhodope Mountains. His singing and playing are said to have been so powerful that they tamed wild animals and moved the rocks and the trees. Some versions of the myth relate that rivers changed course or stood still under the powerful influence of his music.

This myth reflects one perspective on the relations between humans and the environment in Bulgaria's Rhodope Mountains. At the dawn of the 21st century, however, conditions in the Rhodope—the land of Orpheus—are rather different.[1] Instead of the magical power of Orpheus's music, current transformations stem from other, similarly powerful influences. Rural villagers struggle to survive as cooperative farms are liquidated, rural assembly workshops are closed, and the national economy stagnates. Agricultural land and forests have been restored to their former, presocialist owners, and the exploitation of natural resources plays a key role in their postsocialist livelihood strategies. Meanwhile, with the fall of the Iron Curtain, environmental management in former socialist-bloc countries has received considerable attention. In particular, concern about biodiversity conservation as a global priority—key to the story told here—has been imported into Bulgaria in the form of Western-supported projects (World Bank 1994).

Intensified nature conservation efforts are part of a larger global trend in which governments worldwide are being encouraged to protect their biological diversity (IUCN 1994a, 1994b; McNeely 1993; Miller 1984). In Eastern Europe, however, these efforts are substantially influenced by the social, political, and economic changes of the postsocialist period.[2] Conservation challenges facing the region consequently include land reprivatization; changes in agriculture, forestry, and other rural land-use patterns as the economy is reorganized; and limited government resources for environmental programs (Brown and Mitchell 1994; IUCN 1994b, 1995; Meine 1994). Substantial Western concern about the region's environment has been accompanied by the provision of financial and technical support for environmental projects (e.g., Meine 1994; USGAO 1994; World Bank 1994), and a large and diverse body of nongovernmental organizations (NGOs) has emerged to become involved in environmental issues, including nature conservation, at the national, regional, and local levels (see Jancar-Webster 1993, 1998; Pickles and the Bourgas Group 1993; Vari and Tamas 1993).

The countries of the former socialist bloc are now in the midst of social, economic, and political changes moving away from the state socialist system and toward something that is presumed to be more socially just, politically democratic, and economically market-oriented. In Bulgaria, the changes since 1989 have included land restitution, privatization of state enterprises, political pluralism, increased contact with and attention from the West, and less party-state involvement in everyday life. The so-called transition is taking a longer and more tortuous path than initially expected, however, and there is no consensus about the shape of things to come (Giordano 1993; Sampson 1995; Tong 1995). In many countries in the region, wages, living standards, social service expenditures, and production have plummeted, while inflation, unemployment, crime, and prices have risen substantially. Some countries have also seen the return of socialist parties as viable political forces. Sampson (1995) described this context as one characterized by uncertainty and randomness, in which "all things are possible, nothing is certain," and Burawoy and Verdery write of a radical shift in "the rules of the game, the parameters within which actors pursue their daily routines" (1999:2). Fifteen years after the crumbling of communist regimes in Eastern Europe, it is now possible through this Bulgarian case study to assess the real impact of "transformation," not only for human social relations, but also for institutions and natural resources.

Why Bulgaria?[3] Why the Rhodope Mountains? While Bulgaria has not escaped the environmental contamination and natural resource degradation associated with state socialism in Eastern Europe (see Carter and Turnock 1993; Simons 1994; World Bank 1994), substantial areas of relatively unspoiled mountain landscape remain, and Bulgaria ranks among the more biologically diverse countries in Europe (Baker and Baumgartl 1998; Meine 1994). Nature protection efforts are not new to the country, but environmental concerns have received increased attention from the Bulgarian government, NGOs, and bilateral donors in the last decade. The Rhodope Mountains, which are located along Bulgaria's southern frontier with Greece, were selected as the specific focus of this book for several reasons. The first is their high level of biodiversity—that is, the number of species and the richness in rare, endemic, and relic species—and consequent identification as a conservation priority by reports such as Bulgaria's national biodiversity conservation strategy (Meine 1994). The second reason is that, although the Rhodope Mountains are the only mountain range in Bulgaria without a large-scale protected area (e.g., national park or protected landscape), environmental NGOs at local, national, and international levels have been involved in a variety of small-scale conservation and sustainable development projects in the region during the 1990s. Third, the Rhodope Mountains are the most populated of Bulgaria's mountains, which accentuates the importance of local communities in conservation and sustainable development efforts. Despite the changes of the last century, Rhodope residents continue to depend on the local landscape for small-scale agropastoral production and the exploitation of resources such as medicinal plants and firewood. Indeed, such activities may be of increased importance for villagers' subsistence under the unstable conditions of postsocialism. Yet, it is this same rural landscape that conservationists seek to protect.

The Social Dimensions of Conserving Biological Diversity

The last two decades have witnessed increasing concern about environmental protection on a global scale (French 1995). As part of this concern, governments around the world are being encouraged to inventory their biological diversity and to take steps to preserve it by designating substantial portions of their territories as national parks or other types of protected areas (IUCN 1994a, 1994b; McNeely 1993; Miller 1984). This heightened concern about biodiversity conservation stems from the growth of industrial society and human populations, the resulting demand for natural resources, and the consequent threat to biodiversity. In turn, arguments for preserving biological diversity at the genetic, species, and ecosystem levels are usually couched in utilitarian terms emphasizing the potential usefulness of these resources for humans as well as for life on earth generally (IUCN/UNEP/WWF 1991; McNeely 1994; Wilson 1992:chap.13). For example, Jeffrey McNeely of the World Conservation Union (IUCN) wrote that protected areas "help to maintain the diversity of ecosystems, species, genetic varieties and ecological processes . . . which are vital for the support of all life on Earth and for the improvement of human social and economic conditions" (1994:390). As of 1997, 12,754 sites met the criteria for IUCN's Protected Area Management Categories I-VI, representing an area of about 13 million square kilometers or nearly 8 percent of the globe's land area. In addition, several thousand sites do not meet the IUCN criteria but are nonetheless accorded some type of protected status, thus contributing to conservation objectives (IUCN 1994a, 1998).

Two related factors characterize many recent discussions of biodiversity conservation. The first is recognition of the importance of protecting entire ecosystems, rather than just islands of wilderness in a sea of development (McNeely 1994:390; Wells and Brandon 1992; Western 1993). This line of reasoning recognizes that only a small portion of the planet can reasonably be protected by cordoning it off as national parks or strict nature reserves according to what is sometimes referred to as the United States model, which excludes resident human populations, and that environmental protection efforts must, consequently, extend beyond the borders of protected areas to the remainder of the globe (McNeely 1994). The second factor is growing attention to the social dimensions of biodiversity conservation, which stems from a realization that most parts of the earth have human residents or at least people with historically legitimate claims to the land (McNeely 1994). Earlier conservation efforts focused on the territory within the borders of protected areas, viewed human use or occupation of these lands as incompatible with conservation objectives, and paid little attention to the needs and concerns of people living in or around the lands (Allin 1990; Machlis and Tichnell 1985). Now one sees books on conservation with titles such as *Resident Peoples and National Parks* (West and Brechin 1991), *Conservation through Cultural Survival* (Stevens and De Lacy 1997), *African Wildlife and Livelihoods* (Hulme and Murphree 2001), and *Communities and the Environment* (Agrawal and Gibson 2001).

As this change in philosophy now recognizes, local residents affect conservation efforts and are affected by them. In cases such as Rara National Park in Nepal and Kidepo Valley National Park in Uganda, people were relocated in conjunction with the creation of the parks, and the associated human costs were substantial (Heinen and Kattel 1992; Turnbull 1972). Alternatively, people living near areas managed for conservation may be subjected to policies that restrict land use and access to important resources such as agricultural land, pastures, game, and forest products; policies that ignore traditional resource management institutions; or both (Brower 1991; Hitchcock 1995; Homewood and Rodgers 1991; McCabe et al. 1992). Local people may see the goal of protecting wildlife as an outright threat to their survival. For example, Bonner (1993) cites examples from Africa in which the animals protected by a park trampled fields, thereby destroying crops, or stole domestic animals or small children. Similarly, a survey in the Annapurna Conservation Area in Nepal found that local residents felt threatened by the local population of snow leopards (Oli et al. 1994). Such conditions may lead to conflicts between conservation efforts and local people, which can have negative consequences for protecting the integrity of ecosystems when residents fail to support or even actively resist park policies by poaching, logging, farming, or otherwise "illegally" using the habitat (Bonner 1993; Brower 1991; Ghimire 1994).

The relationships between conservation activities and local populations have been described in both policy documents and scholarly literature. Within the social sciences, attention to these issues has focused primarily on Africa (e.g., Grimm and Byers 1994; Hulme and Murphree 2001; Marcus 2001; McCabe et al. 1992; Neumann 1998), Latin America (e.g., Nygren 2000; Staver et al. 1994), and Asia (e.g., Ghimire 1994; McNeely 1999; Mehta and

Kellert 1998; Stevens 1993; Vandergeest 1996), with little attention to Eastern Europe. This research indicates that conservation efforts must extend beyond the borders of protected areas and address the needs and concerns of local people if they are to succeed in protecting the world's biological heritage. It emphasizes the importance of understanding local environmental knowledge, social organization, livelihood strategies, and the institutional mechanisms for managing resources and overcoming resource shortages. It also suggests that there are no one-size-fits-all solutions and that human–environment relations are complicated (also Agrawal and Gibson 2001; Brown and Wyckoff-Baird 1992; IUCN/UNEP/WWF 1991; Little 1994; Little and Horowitz 1987; McNeely 1994; Murphree 1994; Stevens and De Lacy 1997; Wells and Brandon 1992; West and Brechin 1991; Western et al. 1994; Zerner 2000). Based on studies of environmental risk perception and action in Madagascar and Brazil, for example, Kottak and Costa (1993) concluded that people will not act to preserve the environment if they perceive it is not at risk. They also suggest that international conservation strategies must be implemented in the short run and in local communities and that these strategies must address the issue of local populations' access to natural resources. Similarly, the debate generated by Hardin's 1968 article on "The Tragedy of the Commons" has demonstrated that access to and control over land and other strategically important resources such as forests, pastures, and fish stocks can influence how those resources are used and managed, albeit not always in predictable ways (see Feeny et al. 1990; McCay 1992; McCay and Acheson 1987; Netting 1981; Ostrom 1990; Peters 1994).

Orlove and Brush (1996) suggest several contributions that anthropology in particular can make to an understanding of biodiversity conservation issues. One is to describe the relationships and interactions between human communities and the plant and animal populations in a given landscape. These might include discussions of crop selection, forest resource exploitation, herding, and hunting. Another is to articulate the relationships between local people and the government agencies and organizations involved in promoting biodiversity conservation at different levels. Third, anthropologists can identify the interests in and concerns about a given environment or landscape by these interested parties. Taken together, their contributions, and also that of this book, is to clarify the differing concerns and definitions of biodiversity held by "local" people and conservation advocates.

Nongovernmental Organizations, the Environment, and Nature Conservation

An important influence on current conservation efforts, including those in Eastern Europe, is the worldwide explosion in the number of NGOs involved in environmental protection and development projects. Indeed, international environmental NGOs sometimes play influential roles in coordinating environmental activism in ways that transcend national borders (Cellarius 1998; Princen and Finger 1994). The term "nongovernmental organization" refers to a large and diverse body of private organizations that exist in the space between government agencies and the private, for-profit commercial sector. Examples of groups that potentially fall under this broad definition include trade unions, religious organizations, foundations, agricultural cooperatives, community associations, sports organizations, environmental advocacy groups, resource management institutions, savings clubs, private relief organizations, and political action groups. Such organizations can exist or operate from the level of the local community or neighborhood up to the international arena (Carroll 1992; Edwards and Hulme 1992, 1996; Livernash 1992). The nongovernmental label can sometimes be a misnomer, as these organizations may receive funding from governments, and in some cases they are initiated or even run by government bodies or employees (Edwards and Hulme 1992; Livernash 1992; Price 1994). Some attention has gone into classifying the different types of organizations: grassroots, intermediary, international, membership support, service delivery, or northern versus southern[4] (e.g., Uphoff 1996; Vakil 1997). Such classifications result in an alphabet soup of acronyms—for example, GOs (grassroots organizations), GSOs (grassroots-support organizations), NNGOs (northern NGOs), SNGOs (southern NGOs), PVOs (private voluntary organizations), CBOs (community-based organizations), and GONGOs (government-organized NGOs). All these acronyms may be confusing, and it is important to look at specific cases to see the actual nature of the groups and the issues addressed.

Nongovernmental organizations play increasingly influential roles in advocating and implementing environmental protection measures, including those intended to conserve biological diversity (Carroll 1992; Doane 2001; Drabek 1987; Ho 2001; Korten 1990; Livernash 1992; Price 1994; Princen and Finger 1994). Writing about Latin America, Price (1994:43) suggests that the growth in the number of environmental NGOs is related to "increased environmental awareness, frustration with governmental institutions, new sources of 'green' funding, and the ideology of sustainable development." Meyer (1995) also points to the importance of funding in discussing the entrepreneurial activities of southern NGOs in obtaining financial support from northern donors. Donor and government interest in supporting the work of NGOs stems, in turn, from assumptions about their effectiveness, efficiency, flexibility, ability to reach the poor, status as something apart from the government, and employment of participatory approaches (Arellano-López and Petras 1994; Carroll 1992; Korten 1990; Tendler 1982). The abilities to reach rural communities and to promote local participation are especially important for projects that seek to address the social and cultural dimensions of conservation (see Brown and Wyckoff-Baird 1992; McNeely 1994; Wells and Brandon 1992).

Of particular interest for the material presented here are assumptions about the ability of NGOs to act as linking agents in conservation activities. Carroll (1992) suggests that intermediary NGOs—a term used to describe national or regional groups—can play substantial roles in supporting or energizing local groups and in providing links to other political and financial powers (see also Fisher 1996 on grassroots-support organizations). Similarly, Princen and Finger (1994) argue that environmental NGOs operating in the international arena play important roles as independent bargainers and as agents of social learning, linking biophysical conditions to the political realm at both the local and global levels. In this scenario, intermediary organizations may also play a part in a chain linking the global with the local. Finally, international NGOs may provide critical technical support and funding to local and intermediary organizations involved in conservation (Murphree 1994).

Like efforts to integrate conservation and local community

concerns, NGOs are receiving substantial attention from the scholarly and policy evaluation communities for their involvement in environment and development issues (Carroll 1992; Drabek 1987; Edwards and Hulme 1992, 1996; Korten 1990; Livernash 1992; Price 1994; Princen and Finger 1994). Research on the role of NGOs in conservation and natural resource management suggests that the organizations play a part in facilitating sustainable development, managing natural resources, and building local capacity. Several factors can impede their effectiveness, however, including conflicts between the NGOs' need for autonomy and governments' need for control, problems with scaling up projects from single communities to regional or nationwide levels, and confusion over the appropriate roles for NGOs. Some authors also call into question the extent to which NGOs are truly participatory (Edwards and Hulme 1992, 1996; Grimm and Byers 1994; Price 1994; Thomas-Slayer 1992; Vivian 1994; Wells and Brandon 1992).

Despite the scholarly, governmental, and donor attention to NGOs, good local-level case studies and analyses of these organizations, based on firsthand field research, are limited, as are the contributions of anthropologists to this literature to date. Much of the research on these organizations, for example, is based on rapid and superficial surveys, which do not address in depth important issues, such as the relationships between NGOs and their intended beneficiaries or local communities, or the relationships between these organizations and the power centers in their countries (Carroll 1992:22; Fisher 1997). In a review article, anthropologist William Fisher described the literature on NGOs in the following fashion: "This literature as a whole is based more on faith than fact: There are relatively few detailed studies of what is happening in particular places or within specific organizations, few analyses of the impact of NGO practices on relations of power among individuals, communities, and the state, and little attention to the discourse within which NGOs are presented as the solution to problems of welfare service delivery, development, and democratization" (1997:441). He goes on to suggest that anthropologists can contribute to an understanding of "how complex sets of relationships among various kinds of associations, the agencies and agents of the state, and individuals and communities have had an impact in specific locales at specific times" (Fisher 1997:442; see also Lewis 1999).

The scarcity of good local-level analyses of environmental NGOs—by anthropologists or others—is particularly acute in postsocialist Eastern Europe. In Bulgaria, as elsewhere in the region, environmental concerns and environmental NGOs played roles in the political changes of 1989. The number of NGOs has increased considerably since the late 1980s; however, environmental concerns have fallen from the top of the political agenda as the deepening economic crisis captures the public's attention (Baumgartl 1993; Crampton 1990; Fisher 1993; Genov 1993; Georgieva 1993; Pickles and the Bourgas Group 1993). Although NGOs in the region have received considerable scholarly as well as donor attention (see Desai and Snavely 1998; Fisher 1993; Jancar-Webster 1993, 1998; Nikolov 1999; Pickles the Bourgas Group 1993; Pickvance 1998; Snavely and Desai 1995; Vari and Tamas 1993), much of the knowledge produced is predominantly descriptive, based on limited fieldwork; it is also focused on issues of environmental degradation and politics, to the exclusion of the conservation activities and NGO-community relations addressed here.

The Ethnography of Socialism and Postsocialism

Coverage of Eastern European countries in anthropological literature by Western-trained scholars is limited when compared to other parts of the world, although the region has received increased attention in the last decade.[5] One reason for this relative dearth of literature is that this region as a part of Europe is sometimes viewed as too close to home compared to the more exotic locations traditionally studied by anthropologists (Cole 1977; Halpern and Kideckel 1983; Hann 1994; Parman 1998). This was compounded by the difficulties of doing fieldwork there during the socialist era (see Creed 1998; Hann 1987; Kideckel 1993b; Sampson 1995; Silverman 2000; Verdery 1996). Specific conditions varied in time and space. A handful of anthropologists from the United States began work in maverick Romania during the 1970s, although conditions for fieldwork there deteriorated in the 1980s (e.g., Kideckel 1993b; Verdery 1996). In the case of Bulgaria, little long-term rural fieldwork took place between that of rural sociologist Irwin Sanders (1949) in the 1930s and anthropologists Gerald Creed (1998) and Deema Kaneff (2004) in the late 1980s. Even so, Creed (1998:22–23) writes that he had begun to wonder if he would ever get to the countryside when he was finally assigned to a village in 1987 (see also Silverman 2000). Canadian anthropologist Asen Balikci (personal communication, 1995) related that he was not permitted to do fieldwork in his native Bulgaria in the 1960s, although he did briefly conduct fieldwork in neighboring Macedonia. A few other social scientists conducted some work in rural Bulgaria during the socialist period; however, the publications resulting from this work have been sparse.[6] Even now that formal constraints have eased, the number of Western scholars who have done long-term ethnographic fieldwork in rural Bulgaria remains extremely limited.[7]

Nevertheless, anthropologists working in Eastern Europe have made important contributions to the understanding of what life there was actually like, both under socialism and during the present postsocialist period.[8] Their microlevel ethnographic research, based on qualitative data and long-term case studies, has shown how the core ideas and institutions of socialism operated from the inside and from the perspective of ordinary people, and also how these ideas and institutions were modified in specific situations, thus illustrating "how the system really works" from the inside— in contrast to stereotyped views and simplistic assumptions from outside observers (Hann 1993c:9). Ledeneva (1998), Smollett (1989), and Verdery (1993) consider the importance of kinship and informal networks in survival strategies under socialism, for example, while Creed (1991, 1995a) discusses the influence of local context on ideological matters and the continuing influence of agriculture on industrial enterprises. Creed's 1998 book on Bulgaria describes the interactions between socialist policies and structures and the actions of villagers, a process that he calls "domesticating revolution." While winning concessions from the socialist system made life more tolerable for villagers, it also produced a complex and intertwined structure, so that changes in one place produced unexpected consequences elsewhere. Taken together, this ethnographic research also points out differences between the various countries in the region—something that Hann (1993a) argues may be even more important in the post-1989 period (see also Pine and Bridger 1998).

Recent literature on postsocialism emphasizes several themes and perspectives to which anthropologists can contribute.[9] The first is getting behind the rhetoric of "transition," with its concepts of "civil society," "privatization," and "democratic institutions," to see what is actually happening to real people in real communities who are trying to survive. Another contribution made by anthropological studies is to examine processes as opposed to focusing on outcomes, and also to observe continuity as well as change. This includes, for example, a consideration of the ways in which preexisting patterns of social and political interaction are important for and reproduced in the process of agricultural decollectivization, as well as the continuing importance of informal social networks for economic survival. Anthropological research can also help in understanding the influence of history, local political and social organization, particular economic conditions, and ecology on the transition process (Burawoy and Verdery 1999; Dunn 1998; Kideckel 1995a; Pine and Bridger 1998; Sampson 1995).

One important aspect of the postsocialist situation, which has been examined by anthropologists and is particularly relevant to the material presented here, is the decollectivization of agriculture, the associated land restitution, and the subsequent reorganization of production (e.g., Creed 1995b, 1998; Hann 1993b; Kaneff 1996, 1998; Kideckel 1993a, 1995a; Verdery 1994, 1999). This research depicts a long and complex process deeply embedded in social relations and the property regimes of socialism. It also suggests that the restitution process has broad social and cultural implications, including the reinforcement of an ideology based on kinship, the emergence of an ideology of individual rights, and changes in politics and state power (Kaneff 1998; Verdery 1994). Many communities discussed in these works have seen a return to or the continuation of cooperative production, and the literature discusses how the organization of cooperatives is embedded in local contexts as well as factors contributing to their continuation (Creed 1998; Humphrey 1998; Kaneff 1998; Verdery 1998).

Other issues tackled more recently by anthropologists of the region include civil society, NGOs, and foreign development aid. Of course, concern about these topics is not unique to the postsocialist world (e.g., Comaroff and Comaroff 1999; Markowitz 2001; Monga 1996), but discussions of civil society take on an additional cachet in the region because of how the term has been deployed by Eastern European dissidents before regime changes. This literature tends to take a critical stance toward development assistance programs and uncritical evaluations of a resurgence of civil society (Abramson 1999; Berglund and Harper 2001; Bruno 1998; Creed and Wedel 1997; Hann and Dunn 1996; Hemment 1998, 2000; Junghans 2001; Wedel 1998a; Werner 2000). It demonstrates the importance of local contexts and discusses the impact of aid on elite configurations (Mandel 2002; Sampson 2002). Together, this anthropological literature provides a reality check on the accounts of other disciplines, which can be based primarily on statistics, brief visits, or both.

The Fieldwork Context: A Brief History of Postsocialist Bulgaria

The beginning of the end of the socialist period in Bulgaria is usually dated to the fall from power of communist leader Todor Zhivkov on November 10, 1989, in an internal communist party coup. Until this time, Bulgaria is often described as having been closely tied to the Soviet Union and more intolerant of opposition activities than some of its neighbors to the north and west. Late 1989—a time of political change throughout Eastern Europe— also saw large opposition rallies in Bulgaria calling for the dissolution of parliament and the adoption of a new constitution, along with the formation of an opposition coalition, the Union of Democratic Forces (UDF). Between November 1989 and the following December, much of the totalitarian apparatus was dismantled (Crampton 1997:220), and the first two postsocialist presidents were supported by the anticommunist UDF. The Bulgarian Communist Party renamed itself the Bulgarian Socialist Party (BSP) in early 1990, and subsequent roundtable talks with the opposition led to arrangements for free elections later in the year. Following the June 1990 election of members to the Grand National Assembly—in which the socialists dominated—a new constitution was passed in 1991.

Although numerous political parties emerged in the space opened up in the early 1990s, politics have been until recently polarized between the BSP and the UDF coalition, which did not become a formal party until early 1997. Few other parties have garnered enough votes to meet the 4 percent threshold necessary to send representatives to parliament. Neither camp won an absolute majority in the 1991 regular parliamentary elections, and thus each in turn ruled in unstable coalitions with the Movement for Rights and Freedom, a party often associated with Bulgaria's Turkish minority (the party has survived a legal challenge that it is an ethnic party and thus illegal under the constitution; Turks comprise the largest ethnic minority in Bulgaria, a legacy of its Ottoman past). A BSP-led coalition received a majority in the 1994 parliamentary elections, and a UDF-led coalition won a majority in the 1997 elections. The 1997 elections came about early at least in part due to citizen dissatisfaction with the country's economic situation, as described below (Daskalov 1998; Wyzan 1998).

The political changes of 1989 and subsequent efforts to reorganize the economy have had a major effect on life in Bulgaria. In addition to domestic developments, Bulgaria's economy was hit hard by changes elsewhere in the region, due to its heavy involvement in and reliance on trade through the Council for Mutual Economic Assistance, the socialist-bloc trade alliance. United Nations economic sanctions against Serbia and Montenegro, with the consequent loss of Bulgaria's main transport corridor to Western Europe, have also had a negative impact on the nation's economy (Crampton 1997:231). Although privatization of industrial enterprises has proceeded slowly, other parts of the economy have liberalized more rapidly. The liquidation of cooperative farms and restitution of agricultural land, for example, have taken place at a faster rate, albeit neither as fast nor as smoothly as some would like. City streets are full of private vendors and small shops. The average Bulgarian cannot necessarily afford the increasingly available supply of Western-produced goods for sale, however, and the vendors may be educated professionals who lost their jobs in the post-1989 period. Price controls on many products were relaxed in February 1991, and prices rapidly doubled (Miller 1997:65). Liberalization of fuel, electricity, and agricultural product prices began in 1992 (Tzanov and Vaughan-Whitehead 1997). Relaxation of exchange rate controls also began in early 1991, and the currency was allowed to float. The exchange rate between the Bulgarian lev (plural, leva) and the U.S. dollar increased from 2.8 Bulgarian leva to the U.S. dollar in January 1991 to 23.9 leva to the dollar in January 1992 (Miller 1997). As part of this restructuring, the Bulgarian government has entered into agreements with those Western organizations often associated with "development" efforts in the so-called Third World, such as the World Bank, the International Monetary Fund, the London Club of commercial creditors, and the European Bank for Reconstruction and Development, as well as bilateral agreements with individual countries (Wyzan 1998). In general, the economic condition of Bulgaria over the last decade has been characterized by periods of relative stability—albeit at a low level for the average Bulgarian—punctuated by crisis.

By most accounts, the worst crisis thus far in the postsocialist period took place in late 1996 and early 1997, while I was in the field. People expressed their views of the economic situation of the time with one-liners such as the following:

> They say that life will be difficult for a few more years . . . and then we will get used to it.

> Democracy in Bulgaria means the freedom to do what you want and to eat what you have.

The first phrase came from a humor-loving villager; the second is from the New Year's Day 1998 edition of a popular, satirical Bulgarian television show. The latter quip refers to the fact that many Bulgarians do not have much to eat given the high prices and low levels of most wages and pensions. Both sayings reflect the poor economic conditions in Bulgaria in recent years and growing disillusionment about the slow rate of economic improvements. The freedom to do what you want is sometimes equated with anarchy, a term used to describe conditions in which "mafia" groups are seen to control the economy and perhaps the government as well, in which crime is rampant, and in which the police may be considered to be at best useless, and at worst involved in what is perceived as a rapidly escalating level of lawlessness (e.g., Daskalov 1998:25).

Conditions were bad when I arrived in Bulgaria for my fieldwork in mid-1996, in the wake of numerous bank failures and substantial currency devaluation earlier that year. They soon were to get much worse. Fourteen of the country's 27 banks failed in 1996, including two of the largest, and the currency lost half its value relative to the U.S. dollar that spring. At the time of my arrival in July, the exchange rate was nearly 200 leva to the dollar, or about three times the rate of 65 to 70 leva to the dollar that existed during my previous visit in 1995. By Christmas 1996, the exchange rate approached 500 leva. A month later, in late January 1997, it was more than 650 leva. At the height of the currency crisis in mid-February 1997—when the currency seemed to go into free-fall and there was near hyperinflation—the exchange rate reached 3,000 leva to the dollar. Inflation was so rapid that wages and pensions could not keep up with the price increases. The government was not prepared to print money in larger denominations, so one left currency exchange offices with pockets full of bills, even after changing only a modest sum. For several weeks I was unable to mail letters from my village field site to the United States, because postage prices had increased but stamps had not been issued in sufficiently large denominations.

Many Bulgarians I spoke with described the economic conditions in the first few months of 1997 as the worst that they had ever experienced. Conditions were worse than during "the war," some said, even worse than the winter of 1990–1991, when food distribution systems had broken down and there were frequent water and power cuts. Not only did people have no money, there was also nothing to buy. During the worst of the crisis, many food stores closed in response to the hyperinflation and speculative nature of the market. Those that remained open had little of use on their shelves. One Bulgarian colleague told me how glad she was to have a grandmother (baba) at home, because the grandmother insisted on keeping the pantry well stocked with sugar, oil, flour, and home-canned produce. Major newspapers ran articles on how to make your own cheese and soap. Women passed around a recipe for "crisis cake," which could be made with the staples of flour, oil, and jam found in most Bulgarian pantries and did not require the unobtainable and unaffordable sugar, eggs, and butter. The president went on prime-time television one evening to explain what was being done to secure a wheat supply for the country in order to avoid the shortage of bread that had occurred the year before. Thank goodness I do not smoke or drink my tea sweetened, as both cigarettes and sugar were scarce in the village where I lived at the time and the nearby municipal center. (These shortages did cause distress for other members of the household in which I was living, particularly the smokers.) Meanwhile, I was more affected by a severe shortage of gasoline, which made long distance travel an "iffy" proposition since bus runs were canceled and gas stations were closed.

Prices, goods shortages, and inflation were the topics of discussion nearly everywhere I went—on the train, in offices and homes, waiting for the bus, and standing in line to buy bread. Sitting in a municipal government office one day in late January 1997, for example, I listened to the office workers calculate how much their monthly salaries were worth according to the day's exchange rate for the U.S. dollar—about $6. With a monthly pension, one retired friend in the city told me a few days later, she could buy one chicken and some rice, although the meat might be forgone in order to buy bread for a few more days. People with hard currency only changed as much money in a day as they needed for that day's expenses. Some of those with only Bulgarian currency spent their pay as soon as it was received, since it was better to have food and toilet paper than rapid devaluing leva. The education system was affected as well. Many public schools were closed for what were called "firewood vacations," when the government could not afford to heat classrooms.

At about the same time, there was also a political crisis. Beginning in December 1996, opposition forces organized protests in many towns against the sitting socialist government. The socialists were blamed for having brought Bulgaria to the brink of economic collapse. Protesters stopped cars, buses, trains, and taxis; and taxi drivers themselves went on strike, further complicating the travel situation. Students in Sofia blocked traffic, and in January there were daily protest marches in the Bulgarian capital. These marches, which included opposition political figures as well as representatives of the Bulgarian Orthodox Church, ended in front of the Alexander Nevski Cathedral, where marchers were joined by many others. One account estimated the crowd at 20,000 to 30,000 people (Open Media Research Institute 1997). Protesters listened to speeches and chanted against "red [i.e., communist] garbage." Despite the ongoing economic crisis, there was a sense of excitement and hope for a better future among the participants in the protests. Eventually, the socialist government, only two years into its five-year mandate, failed to form a government as was required following the recent presidential elections. A caretaker government was appointed by the newly elected UDF president, and early parliamentary elections were called. In these elections, held in April 1997, an opposition coalition of the Union of Democratic Forces and the People's Union gained an absolute majority of seats in the National Assembly (Bell 1998:332).

Economic conditions have since stabilized, wages and pensions have increased somewhat, and those stores that reopened have long been restocked with an ever-expanding—but not necessarily affordable—array of goods. In July 1997, a currency board was created at the insistence of international creditors, and the Bulgarian currency was fixed to the German mark.[10] Although the acute crisis is over, life continues to be difficult for many Bulgarians. For some, it is a daily struggle to buy bread and pay utility bills. Even before the crisis, the average Bulgarian household spent about half its income on food, compared to average food expenditures of less than 30 percent of household income in Western countries, and an estimated 55 percent of Bulgarian household expenditures in 1997 were for food (Tzanov and Vaughan-Whitehead 1997:116; National Statistical Institute 1997:7, 1998:5). During the winter of 1997–1998, many people on fixed incomes had their central heating restricted to a few rooms or turned off entirely because of fears that their utility bills would exceed their income levels. A colleague related one day that the salary from her job in a government research institute would be insufficient to cover the monthly heating bill for the family apartment, although she and her husband both had second jobs so that additional income sources were available for paying the bill. A copy of the London newspaper the *Guardian*, which I picked up on a trip to England in January 1998, included an emergency appeal to readers asking for money to help heat Bulgarian orphanages during the winter and feed their children. In Bulgaria, meanwhile, a widely publicized campaign asked people to buy glossy postcards for a price equal to that of a loaf of bread, with the proceeds supposedly going to feeding these same orphans. Here too is evidence of the disillusionment and lack of trust that permeates much of Bulgarian society today. At a friend's house in Sofia one day, a retired professional pointed unhappily to two or three postcards sitting on a kitchen shelf and told me that less than half the money raised had actually reached the orphanages after deductions for printing costs and the advertising campaign. She clearly felt misled.

Some of my young Bulgarian friends, who previously were determined to stay in Bulgaria in the belief that things would get better, now are beginning to lose hope and to think about leaving for Western Europe or America. Only time will tell how many more will leave and what this trend means for the country's future.[11]

Citizen dissatisfaction with the governments of the last decade and the country's continuing economic problems is reflected in recent political events. In April 2001, Bulgaria's former king, who had lived in exile since the communist takeover in the mid-1940s, returned to the country and established a new political party—the National Movement Simeon the Second. During my fieldwork that spring, popular discourse suggested that many people would "vote for the king,"[12] as they put it, in the upcoming parliamentary elections. He was often described as their "last hope," as they had lost faith in the ability of both the reform socialists and the anticommunists—the two parties who have dominated Bulgarian politics in the postsocialist period—to solve the country's numerous economic and political problems. And this is indeed what happened. One half of the seats in the June 2001 election were won by the former king's party, not bad for a party that had officially existed for only a month or two, and he was subsequently named the country's new prime minister. As such, he is the first former Eastern European monarch to return to formal power. By fall the political winds had changed again, however. In presidential elections in November 2001, a socialist party candidate defeated the seemingly popular incumbent who was backed by the new prime minister and also appeared to have escaped the corruption allegations levied at many Bulgarian politicians.[13] It seems that the former king's popularity was not much help to the incumbent, or perhaps his popularity had waned in the months since his (party's) election. Some now question whether the former king's government will survive its mandate, which few postsocialist Bulgarian governments have done.

These, in short, were the conditions under which the villagers I lived among were trying to make ends meet, and under which environmental NGOs and government officials were promoting sustainable development and biodiversity conservation. From some level of economic and social security during the socialist period, most Bulgarians have been thrown into a period of substantial uncertainty and economic hardship. Given the relatively short duration of the hyperinflation, it is difficult to separate it from the cumulative impact of the adverse conditions of the postsocialist period. While some observers talk about a "transition period" in the region, many anthropologists are reluctant to use this term because it implies that we know where these societies are "transitioning to" (e.g., Burawoy and Verdery 1999; Hann 1994:231; Pine and Bridger 1998; Verdery 1991). They tend to suggest instead paying attention to what is happening on the ground and to real people in the here and now (e.g., Kligman and Verdery 1999).

Methods and Data Collection

Altogether, I conducted two-and-one-half years of research in Bulgaria between 1995 and 2002. The most concentrated period of fieldwork took place between July 1996 and March 1998, with several shorter stays in 1995, 2000, 2001, and 2002. The research has been multi-sited in that, rather than working in a single local community, it moves across levels and between different kinds of communities in tracing relationships between people, organizations, and the state and flows of ideas, people, and funding (Marcus 1995; cf. Wedel 1998a). During 1997, I was based in one Rhodope village and concentrated on the village case study, while my work during the second half of 1996 and in early 1998 was based out of Sofia and focused on NGOs and various programs and projects related to biodiversity and resource use.[14] Throughout this time, however, I traveled periodically between these two settings, as well as to other parts of Bulgaria identified as having conservation value. Language study also took up significant time during 1996. Finally, I traveled briefly to England and Switzerland in January 1998 to visit the headquarters of some of the foreign national and international conservation organizations active in Bulgaria.

During my year in the village, I lived with a family in many ways typical: now heavily involved in private household agriculture, their socialist-era experience included skilled employment in a variety of nonagricultural settings along with their personal-plot production and occasional work on the cooperative farm. My fiftysomething landlord, his wife, and their thirtysomething daughters were of great help in orienting me to the community, and through them I met many of their friends and relatives. My circle of acquaintances extended beyond that of my village hosts as I frequently wandered up and down the main street—an important social space—greeting people as I went and often stopping to talk. Another way of meeting people was through participation in numerous village activities, either in the company of or having been sent by people I already knew. Most of the time I was on my own, though occasionally I asked a bilingual university student to join me for a few days. (Two young women from Sofia and one from Plovdiv served in this capacity over the course of my research, as no one in the village spoke English and they were all too busy for me to rely on a single person as a resource.) Initially they came to the village to help me with questions that I had not been able to sort out on my own, and later, as my Bulgarian improved, they accompanied me on my travels to other parts of the Rhodope.

Qualitative data were collected by observing and participating in everyday life, with particular emphasis on activities involving natural resource use. I herded sheep, cows, and goats; participated in all aspects of potato production, from planting through harvesting and sorting; raked hay; collected mushrooms, herbs, rosehips, and raspberries; went on a firewood collecting expedition; helped plaster an outlying agricultural building; assisted with canning fruit and making jam; learned how to make local cheeses; spent a few days with logging and reforestation workers for the local forest enterprise; and accompanied local hunters on several occasions.[15] While some villagers thought it odd that someone "with two university degrees" wanted to "spend time with livestock," they were gracious and patient in allowing me to accompany them in their activities and in responding to my endless questions. In addition, I drank tea at local coffee bars, sat on streetside benches on warm summer evenings talking to children and grandmothers, hung out at the local hair-dressing salon, went to more funerals than I care to remember, attended village assemblies and festivals, and accompanied villagers on visits to market and administrative centers outside the village. I was also fortunate to make friends with a couple of villagers who enjoyed hiking. While traveling with them across the village territory, they told me about landscape changes that I might not have thought to ask about from a vantage point of the settlement proper. Periodically an assistant and I would search out specific individuals to interview on topics including village history, medicinal plant use, the activities of the local hunting and fishing society, and liquidation of the cooperative farm.

These qualitative activities were combined with several more systematic data-collection efforts. After about five months in the village, I conducted a survey of 120 randomly selected households—approximately one-half the total number in the village—with the assistance of three village teachers. Besides collecting basic demographic information on each individual in the household, questions addressed household participation in farming, collection of natural resources, nonagricultural employment, property ownership, and awareness of natural resource issues. Much of the numerical data presented in chapters 3 and 4 comes from this effort. Follow-up interviews conducted in late 1997 with nine households from that survey serve as the basis for the case studies at the end of chapter 4. Several smaller surveys supplement these data. During the mushroom collecting season, a research assistant and I spent three afternoons interviewing mushroom collectors at three of the largest buying points in the village and in so doing spoke with 37 individuals or groups of collectors. Finally, land ownership and land use data collected on the household survey only provide information on people living in the village, and thus do not show the situation regarding land that has been restored to people living elsewhere. To address this data gap, additional interviews were conducted with people in whose names land was restored or their heirs, regardless of their residence. Forty-two names were randomly selected from a list of 339 people in whose names land had been restored. With the help of villagers I was able to identify and then locate either the individual or at least one heir for 37 of these names (11 percent of the names listed), and my local enumerators plus a teacher in the nearby municipal center interviewed them. (In many cases they were able to interview people living outside the village when they visited the village to see their relatives.) These interviews sought to collect basic data on how land had been divided among the heirs, where each of the heirs lived, and whether and how the land of each heir was being used.

To supplement my firsthand ethnographic and survey data, I collected archival and historical material about the village and the region. This included demographic information from the local administration and government statistical offices. Several days were spent in the regional archives in Smolyan looking for information on historical population trends and agricultural activities. One villager had written a book and two others had written academic theses about the village. Finally, in consultation with Bulgarian scholars of the region, I sought out ethnographic and historical material that had been published about the central Rhodope.

In the absence of a major conservation campaign, NGO activity outside the capital is often scattered and sporadic. Consequently, I decided to combine my intensive village study with a more extensive examination of the environmental NGO community. This effort concentrated, albeit not exclusively, on NGOs involved in biodiversity issues and conservation efforts. Extended interviews were conducted with NGO members and representatives as well as government environmental and natural resource officials at local, regional, and national levels. Throughout this process, I focused on organizations and agencies that were either significant players in Bulgaria generally or that worked specifically in the Rhodope. In association with these interviews, I often collected printed material about the organizations and their activities. Participant observation was employed when the opportunity arose, including attending events sponsored by the organizations, such as a strategic planning seminar, a seminar on increasing public involvement in governmental policymaking in Bulgaria, community meetings held in several Rhodope villages, and an annual meeting. I was in repeated contact with more than a dozen Bulgarian organizations and spoke at length with representatives of four prominent national or international organizations working in Bulgaria or funding the work of Bulgarian environmental organizations. I also investigated the larger institutional context for NGOs and biodiversity conservation in Bulgaria, speaking, for example, with funding agencies and representatives of government-sponsored conservation projects. Taken together, this examination of a locality-based village community and an interest-based environmental community allows me to consider the relations between the two communities as they are linked to the same landscape—the land of Orpheus.

Previewing the Argument

The ethnographic study of one Rhodope village is framed by an examination of NGOs and nature conservation in Bulgaria generally and in the Rhodope Mountains in particular. Chapter 1 sets the stage by describing nature conservation activities in Bulgaria and the involvement of NGOs therein. This historical discussion begins in the early 20th century, continues through the socialist era, and concludes in the 1990s. Chapters 2 through 4 present an ethnographic case study set in a landscape that NGOs and other conservation advocates identify as important from the standpoint of biodiversity conservation. This story of one mountain village provides insights into rural life and the use of the natural resources that conservation efforts aim to protect. Chapter 2 introduces the village and describes resource use during what villagers refer to as "private times," as well as during the socialist era. Chapter 3 presents a detailed description of postsocialist farming, herding, hunting, and gathering, ending with a consideration of the implications of Bulgaria's postsocialist agricultural-land restitution for resource use and general village life. Chapter 4 examines how this resource use fits into overall livelihood strategies of village residents and considers the strengths and weaknesses of these survival strategies. Chapter 5 returns to the theme of NGOs and biodiversity conservation in examining several aspects of NGO activities in rural areas and particularly in the Rhodope. Finally, chapter 6 presents a social biography of organizational life in another Rhodope village and compares it to the case study's village, which appears to lack such associational activity.

In the Land of Orpheus tells the tales of two communities interested in Bulgaria's Rhodope Mountains—rural villagers and conservation advocates—and seeks to understand how the everyday realities of rural residents relate to global concerns about biodiversity. This account provides important microlevel insights into postsocialist conditions, insights that come from long-term fieldwork and are often missing from more superficial accounts. Detailed descriptions are offered of the ways in which mountain residents use natural resources and factors affecting the shape of resource use, on the one hand, and the impact of the national economic and political context on both environmental NGOs and rural communities, on the other, thus setting the stage for this book's conclusions. These conclusions focus on the current and possible future roles of NGOs in conservation efforts, particularly in relation to rural livelihoods and village communities. While Rhodope villagers and conservation advocates each value the landscape differently, there is some potential middle ground for constructive discussion of sustainable resource use in the region. Along with their activities in the policy arena, Bulgarian environmental NGOs play a role in linking global conservation concerns with the local realities of rural communities, although they are no "magic bullets." In the uncertain and dynamic postsocialist context, their efforts are hampered by funding concerns and the fact that they are not always recognized as legitimate players. Finally, looking toward the future of the Rhodope, NGOs are likely to continue to play a role in conservation efforts there, including those involving relations with rural communities.

1

Bulgarian Environmental NGOs and Nature Conservation

A Historical View

Most accounts of Bulgaria's environmental movement begin with mothers' protests over air pollution in the northern Bulgarian town of Ruse in 1988 and the formation of an organization often referred to as the Civil Committee for the Ecological Defense of Ruse. The narratives then move to the establishment of the organization Ecoglasnost in 1989 and its role in the fall of socialism in the country (e.g., Baumgartl 1993; Crampton 1990; Fisher 1993; Jancar-Webster 1993; Pickles and the Bourgas Group 1993; Snavely 1996; Snavely and Desai 1995). Reading these accounts, one is sometimes left with the feeling that the environmental movement and even concern about the environment by individuals and independent groups began in Ruse and Sofia in the late 1980s. Desai and Snavely (1998:35), for example, date the beginning of an environmental movement in Bulgaria to a specific day—September 8, 1987—and Mitsuda and Pashev write that "the environmental movement in Bulgaria sprang to life as a herald of revolutionary social transformation" (1995:107). Some of the authors cited above additionally suggest that this environmental protest can be seen more as a feasible way to challenge the old regime—more acceptable than protest actions based on ethnic or human rights issues—than as an effort to address environmental concerns.

These protests and organizations are undeniably important in the history of the environmental movement in Bulgaria and also are significant for the role they played in the fall of the dictator-ship of Todor Zhivkov and the end of the communist party's po-litical monopoly. It is also not out of the question that some people were attracted to the movement largely as a way to challenge the old regime (Wedel 1998a:113). Yet, the focus on these most visible of events and organizations tells only part of the story. Less atten-tion has been given to other, sometimes less political organiza-tions that also started their activities in the late 1980s, and one rarely sees discussion of environmental concerns in the country dating before this time—despite the fact that the history of or-ganizations with environmental purposes goes back more than a century. Consideration of this history is important to a more thor-ough understanding of environmental activism in postsocialist Bulgaria.

Early Conservation Efforts and Conservation Organizations in Bulgaria

The origins of environmental protection and conservation ac-tivities in Bulgaria are closely tied to the beginnings of the non-governmental environmental community in the country and date back to the late 19th century. The Bulgarian government's first leg-islative measures to protect natural resources came in the decade following independence in 1878, with the passage of laws on fish-ing, hunting, and forests (Rizov 1987). The first organization with a specifically environmental focus was the Bulgarian Nature Re-search Society, established in 1896, and the early 20th century saw the creation of other professional associations with environmen-tal interests—a botanical society, a geographical society, a union of foresters, and so on. In 1928, ten nature protection societies and organizations involved in nature research joined together to form Bulgaria's first organization devoted explicitly to nature conser-vation, the Union for the Protection of Nature (sometimes trans-lated as the Council for the Protection of the Countryside), and the organization was officially registered in 1929.[1]

The Union and its member organizations subsequently played a role in the creation of Bulgaria's first protected areas and pas-sage of the country's first law on nature conservation (Suiuz za zashtita na prirodata 1995). In 1926, some of these organizations, including the Bulgarian Nature Research Society and the Bulgarian Botanical Society, proposed the creation of Bulgaria's first nature reserve, Gorna Elenitsa-Silkosiya, located in the region of Strandja Mountain (Georgiev 1993). This proposal came to fruition in 1933 when Bulgaria established its first two nature reserves, Silkosiya (a shortened version of the originally proposed name) and then, a few days later, Parangalitsa in the Rila Mountains. The Union was also involved in discussions leading to the creation of Bulgaria's first national park, on Vitosha Mountain, in 1934. The people involved in these actions also saw the need for special legislation on protected areas, and the resulting Law for the Protection of Bulgarian Nature was passed in 1936. It specified several different types of protected territories—reserves (originally called *branishta* [protected places]), national parks, natural landmarks, and natural-historical places. The law called for special regimes for the protection and use of these territories, although it did not necessarily change their ownership status. A few more reserves and small national parks were created in the 1930s and early 1940s, including Bistrichko Branishte and Torfeno Branishte on Vitosha Mountain, thereby at an early date beginning the tradition of designating strict nature reserves within national park borders (Georgiev 1993; Peev et al. 1995:85). Thus, between Bulgaria's independence in 1878 and the end of World War II, Bulgaria saw the passage of a nature protection law, the creation of some protected areas, and the adoption of implementing regulations for a few parks. Actual protection was weak, however, and little research had been done on the flora and fauna within protected areas (Georgiev 1993:52–53).

Conservation Developments under Socialist Rule

With the advent of socialist rule in the mid-1940s, most independent nature conservation organizations in Bulgaria either disappeared or lost their autonomy by being incorporated into or otherwise associated with state-controlled organizations. (Some of these organizations, including the Union for the Protection of Nature and the Union of Foresters, subsequently reasserted their independence after 1989 and registered as independent, nongovernmental organizations.) A Committee for Nature Protection was created as a public organization during the 1970s; however, the formation of independent NGOs was discouraged, information about environmental issues was often suppressed, and citizens had little ability to influence decision making on environmental matters (IUCN 1991, 1994a; Pickles and Mikhova 1998). One postsocialist NGO official said that the primary role of these semipublic, semigovernmental organizations was to convince the public that the government's decisions were correct.

After a period of little action during World War II and the following decade, conservation again received attention in the 1960s with the passage of a new law on nature protection and the creation of additional protected areas. The 1967 law on nature protection lists five types of protected areas: nature reserves, national parks, nature sanctuaries, protected sites, and historical sites. It also includes provisions for protecting specific plant and animal species (Georgiev 1993; IUCN 1991). Some smaller reserves and other protected sites were designated in the 1960s and 1970s, and the second Bulgarian national park of any substantial size, Pirin National Park, was established in 1962.

Bulgaria also participated in international programs and activities for biodiversity conservation during the socialist era. For example, a special issue of the magazine *Bulgaria*, dedicated to the European environmental ministers' meeting held in October 1989, lists numerous international environmental forums hosted by Bulgaria between 1984 and 1989, including an international symposium on the protection of nature reserves and their genetic stock and meetings of government experts from Balkan countries on ecological protection issues on the peninsula. Bulgaria signed the Convention for the Protection of the World Cultural and Natural Heritage in 1974 and has nine sites on the World Heritage List, two of which are natural sites. On both counts, this is more than any other country in Central and Eastern Europe (UNESCO World Heritage Centre 1999). Bulgaria is also a party to the Ramsar Convention on Wetlands of International Importance, in force in the country since 1976, and has four Ramsar sites. In addition, individual Bulgarian scientists were members of some commissions of the World Conservation Union (IUCN).

Such actions may have been in part political moves designed to show that Bulgaria was doing something to protect its environment, as much as they were actual commitments to on-the-ground action. A case in point may be the country's participation in the Man and the Biosphere Program of the United Nations Educational, Scientific, and Cultural Organization (UNESCO). As of 1997, Bulgaria had designated 17 biosphere reserves under this program, including four in the Rhodope, and only two countries had more—the much larger United States, with 47 sites, and the Russian Federation, with 19 sites (IUCN 1998). Yet, Bulgaria's sites are all small, ranging in size from 576 to 2,889 hectares (ha), and thus the number of sites and the country's associated third-place ranking may sound more impressive on paper than is warranted in terms of the total area protected. Similar evidence for conservation on paper without necessarily providing protection in practice is seen in the fate of a small national park, Zlatni Pyassatsi (Golden Sands), which had been created in 1943. By the time the enormous socialist-era Black Sea resort of the same name was built, virtually nothing remained of the park (Peev et al. 1995:176). Thus, in this case, economic development interests took higher priority than enforcement of conservation policies. Indeed, it has been observed that Bulgaria had good environmental laws during this period, but they were not necessarily enforced (Friedberg and Zaimov 1998; Koulov 1998:145).

Socialist-Era Environmental Actions and Organizations

Most observers, and the few participants that were on the scene, report that organized environmental action and the accomplishment of concrete activities independent from the party–state and its organs was difficult during the years of socialist rule in Bulgaria. Indeed, Bulgaria is often cited as one of the countries in the region in which independent activities were most difficult to carry out and most strongly discouraged (e.g., Fisher 1993; Verdery 1996; cf. Weiner 1999). To say that nothing happened regarding the environment under those conditions is inaccurate, however. Developments during this period helped set the stage for events in the late 1980s and into the 1990s. A relatively high level of literacy in Bulgaria, including at the college level, meant that there was a scientific community of professional zoologists, botanists, ornithologists, foresters, and so on who collected environmental data through their work in universities and research institutes, and a few of them participated on IUCN commissions. Pickles and Mikhova (1998:107) observe that such research increasingly took place in the 1980s, although the results were rarely available to the wider public. At least some of these individuals thought about the environmental issues raised in their work, even if they could not safely raise their voices in public protest. For example, several researchers who had been involved in an environmental study of the Burgas area in the 1980s—the results of which were suppressed for many years—subsequently played instrumental roles in organizing an environmental movement in that locality in 1989–1990 (Koulov 1998:207). Similarly, the wife of another organization's founder said that she had listened to her scientist–husband talk about his ideas for an environmental organization and its activities "for 20 years" before he was able to legally establish the organization in 1990. Interaction among these individuals also took place on periodic scientific expeditions, which often involved students and invited environmental experts. One participant described these expeditions—on which participants made environmental observations and sometimes wrote reports at the end—as the main way in which people could work together on environmental issues at the time. In the late 1980s, some urban-based environmental specialists took action by establishing feeding stations for birds of prey in the eastern Rhodope Mountains in an effort to support specific populations of rare birds. Although government agencies were occasionally persuaded to provide limited funds for this activity, it was an independent effort by individuals. Thus, while independent environmental action was constrained, some scientific research was carried out during the socialist period, and individuals continued to be trained in and to work in environmentally relevant fields.

In addition, earlier during the socialist period, at least one and perhaps more short-lived attempts were made to form independent environmental organizations. In 1976, students from the biology faculty at Sofia University established the Youth Club for the Conservation of Nature. According to the organization's former president, interest in the group was high, and many people, including students from other universities, soon joined. He described this organization as an Ecoglasnost-prototype organization, an informal group, which ignored Comsomol, the communist youth organization under which it could have organized.

These efforts initially received good reviews in the press. For example, an article in the newspaper *Narodna Mladezh* (National Youth) concluded with the following statement: "One young club begins its activity—an activity that is necessary, infinitely important in the technological era in which we live and which threatens the nature around us. Let us wish it a good journey, let us wish it success" (Bogdanova 1977 [my translation]; see also Dimitrov 1976). Although organization leaders were never directly told that their group was illegal, eventually they were unable to find a place to meet. The government reportedly created a parallel organization—a student's environmental club at the Higher Forestry Institute[2]—to divert attention and support from the independent student club, which was told that there was no need for their organization. The organization eventually disbanded in 1977, with its only concrete activity being a tree-planting effort. Its former president went on to hold a leadership position with a large Bulgarian environmental NGO in the postsocialist period, however, and several others involved in that earlier effort also participate in present-day organizations. This example shows pre-1989 environmental concern by individuals who subsequently joined NGOs when this was a viable option. An undated information sheet from the organization Green Patrols tells a similar story of starting environmental activities that were repeatedly squelched by the communist authorities.

Some of today's independent groups have their organizational foundations in the officially recognized organizations of the 1970s and 1980s. One example is the student organization at the Higher Forestry Institute, which continues in the postsocialist period as an independent NGO. Another example is the University Rescue Squad, which was established as an NGO in 1990. It has roots in an earlier organization based at Sofia University that participated in mountaineering and scientific expeditions with official sanction but little or no government financial support during the 1970s and 1980s. The organization could no longer afford to travel after 1989 because of increased costs and poor economic conditions, but some individuals involved felt that they had a strong group and wanted to continue. So, they established the University Rescue Squad and initially undertook activities such as mountain rescue and responding to pollution or contamination incidents, which made use of their technical climbing skills and scientific knowledge associated with their positions as students in the sciences. The organization also planned to conduct a summer seminar in 1997 for young people on biodiversity in the Rhodope, which is how I came into contact with it, although funding problems resulted in the seminar's cancellation. This organization's activities have branched out in an unexpected direction, but that subject will be discussed later in this story. First, attention must be directed to recent developments in conservation and the reemergence of a formal nongovernmental environmental sector in Bulgaria.

Bulgarian Nature Conservation in the 1990s

Postsocialist economic and political conditions in Bulgaria have affected conservation efforts both positively and negatively. The environmentally friendly character of the Grand National Assembly is reflected in Article 15 of Bulgaria's 1991 Constitution, which states that the "Republic of Bulgaria ensures the protection and conservation of the environment, the sustenance of animals and the maintenance of their diversity, and the sensible utilization of the country's natural wealth and resources" (translated in Flanz 1992:89). With the relaxation of restrictions on their existence, nongovernmental organizations are again involved in environmental issues in Bulgaria, and foreign governments and other donors are providing financial and technical assistance for biodiversity conservation and other environmental issues. Conservation efforts by the Bulgarian government along with efforts sponsored by other governments and donors are briefly described here, and the remainder of the chapter concerns nongovernmental organizations and the larger context in which they operate.

The focus of government conservation efforts in Bulgaria shifted in the 1980s and 1990s, from relatively small nature reserves to national parks covering larger geographic areas. This parallels a worldwide trend since the 1970s of significant increases in the area of land under such formal protection (e.g., IUCN 1994b:251–252). Recently established parks include Vratchanski Balkan in 1988 (30,130 ha), Central Balkans in 1991 (73,262 ha), Rila in 1992 (107,924 ha), and Strandja in 1995 (116,260 ha) (IUCN 1998). Their establishment has led to a substantial increase in the area accorded some form of protected status in Bulgaria. Thus by 1997, Bulgaria had 12 national parks, 90 nature reserves, and hundreds of other protected natural phenomena and sites (National Statistical Institute 1998:2). Protected areas covered 491,219 ha or about 4.4 percent of Bulgarian territory, about a threefold increase in a decade (Ministry of Environment and Waters 2000: 16).[3] Bulgarian conservationists also take pride in what they describe as "the greatest network of strict reserves in Europe" and the fact that 60 percent of the territory of strict nature reserves is located within the boundaries of national parks, so that the parks serve as buffer zones to the more strictly protected and ecologically important reserves (Peev et al. 1995:85).

The institutional structure for protecting these sites is still developing, and conservation efforts have been hampered by recent changes in the country. But this situation is beginning to change, particularly with foreign technical and financial assistance and the activities of NGOs. The Ministry of the Environment (now the Ministry of Environment and Waters) was established in 1990 as a ministerial-level body with responsibility for environmental issues (Baker and Baumgartl 1998). A National Nature Protection Service (NNPS) was created within the ministry in 1994 as the government unit with primary responsibility for protected areas and biodiversity conservation. The latter was part of an ongoing institutional-strengthening effort, sponsored in part by external donors, to help Bulgaria create its own protected-area management system. While the country has had protected areas for some time, many of them have no management plans, and protected-area oversight has been scattered among many different government bodies, often local forestry authorities and municipalities (Mihova 1998:708–709). A summary of the NGO contributions to Bulgaria's biodiversity strategy, presented in March 1993, includes the statement that "the members of the NGOs report that, in the course of the many visits they have made to protected areas throughout the country, they have never been inspected by anybody. This leads to the conclusion that most of the protected areas exist only on paper" (Mihova 1998:709). As well, the work of the NNPS has been hampered by a lack of resources. In a 1995 interview, the service's director expressed a desire literally to mobilize his employees, by providing them with the vehicles needed to do their jobs. More generally, Baker and Baumgartl (1998) write that the Bulgarian government in many instances lacks the administrative and institutional capacity to tackle environmental problems.

Some changes have taken place on the legislative front, however. Along with the environmentally friendly constitution, the Grand National Assembly passed a new Law on Environmental Protection in 1991 containing strong provisions for environmental impact assessments. (Subsequent amendments weakening these provisions are discussed later in this chapter.) Work on postsocialist versions of other relevant environmental legislation proceeded more slowly (Friedberg and Zaimov 1998; Georgieva and Moore 1997). A new Law on Forests was not passed until late 1997; it accompanied separate legislation to restore private forests to their former owners. Other recent environmental legislation relevant to environmental conservation has included postsocialist laws on protected areas (1998), medicinal plant protection (2000), hunting and game protection (2000), biodiversity (2002), and a new Environmental Protection Act (2002).

The new protected-areas law sets out the categories of protected areas in the country, their purposes, and the conditions for their protection, use, declaration, and management (State Gazette No. 133, 11 November 1998). Reflecting the more open political environment, NGOs were consulted in drafting this legislation, and some provided written comments on the proposed law. Of particular concern for some environmentalists was a distinction made in the legislation between national parks, which are protected areas owned exclusively by the government, and nature parks, which have private or mixed ownership.[4] Under the new legislation, only three of the country's twelve parks—Rila, Pirin, and Central Balkans—are accorded the status of national parks, in part due to this distinction. The others were generally reclassified as nature parks. In both cases, the law requires the development of park management plans within a three-year period, as well as a review of the existing park boundaries.

In a context of limited financial resources, much of the postsocialist activity toward developing a protected-areas system in Bulgaria has come through projects sponsored by foreign governments and other international organizations. In the early 1990s, Bulgaria formulated a strategy for the conservation of biological diversity with assistance from the U.S. Agency for International Development (USAID) and the Biodiversity Support Program, which is a consortium of three U.S.-based environmental organizations—the World Wildlife Fund, the Nature Conservancy, and the World Resources Institute. Five Bulgarian NGOs—the Bulgarian Society for the Protection of Birds, the Bulgarian Society for the Conservation of the Rhodope Mountains, Ecoglasnost-Varna, the Green Balkans, and the Wilderness Fund—participated in setting conservation policy at the national level by contributing papers to and participating in a 1993 workshop from which the strategy developed (Meine 1994, 1998). A national action plan for the conservation of the most important wetlands in Bulgaria was similarly elaborated with the support of the French government and the Ramsar Convention Bureau (Ministry of Environment 1995). Subsequent activities have often been guided by Bulgaria's national priorities identified in these documents. The British Know-How Fund sponsored a three-year project called the Pirin Rila Eco and Sustainable Tourism Project (the PREST Project), which focused on small business development and environmental education in the mountains of the country's southwest corner. Other small, foreign-sponsored projects have included PHARE[5] funding for a visitor center in Rila National Park and a coffee-table book on Bulgarian nature; and UNESCO funding for monitoring and public information activities at the Sreberna Reserve, a wetland site near the Danube River that is both a Ramsar site and a World Heritage site. The two largest and most prominent projects are sponsored by the governments of the United States and Switzerland.

Besides U.S. government support for the biodiversity strategy, the United States' early 1990s involvement in Bulgarian conservation included construction of a visitor center for Vitosha National Park. More recently, funding has been provided through USAID for what is often referred to as the GEF Biodiversity Project.[6] On the Bulgarian side, this project works most closely with the environmental ministry. The main project goals are to strengthen the park system through policy and institutional development (e.g., legislation and training of park professionals); to develop management plans for the Central Balkans and Rila national parks, with an eye toward their being models for other park management plans; to increase public awareness regarding parks and nature protection; and to develop financial mechanisms to support conservation in the long run. The follow-up Biodiversity

Conservation and Economic Growth Project also includes a component for income generation along park borderlands through such things as ecotourism and enterprise development. This project does not directly involve environmental NGOs, although in principle it can contract with the groups for the provision of particular products, such as training sessions or reports.

The second large foreign-sponsored conservation project, the Bulgarian-Swiss Biodiversity Conservation Program (BSBCP), is funded through the Swiss Agency for Development and Cooperation and based on an agreement between the Bulgarian Ministry of the Environment and the Swiss agency.[7] Like the GEF Biodiversity Project, its main objective is to contribute to biodiversity conservation in Bulgaria through both capacity building and actions on the ground. Project implementation and oversight has taken place through three Swiss NGOs. The Swiss League for Nature Protection is the lead organization, and it works in conjunction with the Swiss Association for the Protection of Birds and the World Conservation Union (the latter was only involved in the first phase). The Swiss program works primarily on Black Sea Coast wetlands, the Central Balkans National Park, coastal Dobrudja, the eastern Rhodope Mountains, and Strandja Mountain. Development of a management plan for Pirin National Park was added in the program's third phase. The Swiss NGOs work in partnership with Bulgarian environmental organizations in some projects under this program; in others, they work with the Bulgarian government.

The situation in postsocialist Bulgaria has influenced both the duration of the Bulgarian-Swiss program and the way its money is managed. The program was initially intended to have only one phase, between 1994 and 1997; however, program sustainability was not achieved in this time frame due to the economic crisis and disorganization of the government sector in Bulgaria. A second phase, from 1998 to 2000, concentrated on developing and implementing management plans along with establishing information and training centers in hopes of having these conservation efforts continue after Swiss financial support stops, and now a third phase is expected to run through 2004. Conditions in Bulgaria have also affected the way the program's money is administered. The original plan was to route the funding through the Ministry of the Environment. The Swiss authorities decided that this was not feasible given a lack of appropriate accounting and financial management structures within the ministry, and consequently the funds are channeled through a program administrator in the BSBCP office. Because it was necessary for the money to pass through a legal structure in Bulgaria, BSBCP was registered as a foundation under Bulgarian law—the same law under which NGOs are registered—with representatives of the Swiss NGOs serving as officers of the Bulgarian NGO. Thus, a lack of financial management capacity by the Bulgarian government prompted the creation of a new NGO to manage the money for this government-sponsored project.

The Reemergence of a Formal Nongovernmental Environmental Sector

The Bulgarian Society for the Protection of Birds is often as identified the first independent environmental NGO to officially register in Bulgaria in the 1980s, doing so in December 1988. This organization was created with the goal of achieving progress toward environmental protection with concrete actions on the ground, which its founders felt were lacking in the government's approach toward conservation at the time. Some of its founders had been involved in the above-described efforts to feed birds of prey in the eastern Rhodope Mountains in the late 1980s thus supporting certain bird populations, in contrast to the Bulgarian government's strategy of passing laws that were not enforced and participating in international meetings.[8] The best-known environmental group in the late 1980s was Ecoglasnost, which formally registered in 1989 and has numerous local branches and namesake organizations around the country. The goals of Ecoglasnost included free access to information, the protection of human health and safety, and political change toward a democratic system of government (Pickles and the Bourgas Group 1993:173). The organization's name reflects an interest in the policy of *glasnost* or openness, which was being promoted by the Soviet leadership at that time, applied to the field of ecology.

During the late 1980s, other groups started to coalesce as well. For example, Green Balkans formed partially in response to a government campaign to poison the burgeoning vole population in Bulgaria during winter 1988–1989, a campaign that resulted in the deaths of several thousand birds, including specimens of rare and threatened species, along with the targeted population of voles. In 1988 the then-unofficial group held a protest demonstration in Plovdiv, Bulgaria's second largest city, with participants carrying the bodies of several poisoned birds. It also produced video footage of the dead birds, which was reportedly shown on television as a background to a government official stating that there was "no problem"—that there were no dead birds. The activist who told me about this said that people might not remember the name of their organization, but they remember seeing the dead birds on TV. This organization tried unsuccessfully to register in September 1989 and succeeded a few months later after the political changes.

With the fall of the former government regime in November 1989, the conditions were considerably liberalized for the reemergence of a formal nongovernmental environmental community in Bulgaria. From a handful of independent organizations in 1989, their numbers have grown substantially such that there are currently 100 to 150 environmental NGOs in Bulgaria, many of them created in the period 1989–1991. These figures are ballpark at best, which is important to recognize because growth in the numbers of such organizations is often cited as a measure of the development of civil society in the region (e.g., Massam and Earl-Goulet 1997; Snavely and Desai 1995). It is beyond the scope of this discussion to investigate in detail this alleged connection (see Cellarius and Staddon 2002; Hemment 1998; Wedel 1998a); however, a few words about numbers are in order because of the attention they receive.

Several factors make it difficult to determine the exact number of active environmental organizations in Bulgaria today. Until recently there was no central government registry of NGOs, with such organizations simply being registered in district courts around the country (Kyutchukov 1995). The 2000 NGO law establishes such a registry, but organizations have three years to re-register, and the central registry only applies to those organizations that are designated as public benefit groups.[9] Another question is which organizations should be considered as environmental groups. For example, it is not clear how to appropriately classify a local economic-development or sustainable-development organization, which focuses its efforts on economic issues but whose bylaws include language regarding sustainable resource use; or a professional association in an environmental field such as botany or forestry; or an organization that is registered under the NGO law, but which operates more like an environmental consulting firm—using its status as a legal person to compete for project funds that pay its employees—than as a popular, membership-based organization. Groups falling into each of these categories could potentially work on environmental issues, but they may or may not be counted, depending on who is doing the counting and for what purpose. Beyond this question of definition, there may be informal groups that are not officially registered, or registered groups that are temporarily or permanently inactive, due to leadership lapses or a lack of financial resources. In 1997, one observer of this community estimated that there were about 150 registered environmental organizations in Bulgaria but only about one-third of the groups were active. Counting the number of environmental NGOs in Bulgaria is further complicated by the question of how to treat branches or other subdivisions of the organizations—that is, as separate groups or parts of one main group. As will be seen below, this is an issue about which the NGOs themselves are sometimes unsure, but the important point is that counting these organizations can be difficult and subject to decisions about what should be counted.

Another difficulty occasioned by the fixation on numbers, as a measure of civil society development, for example, arises with the participation of prominent activists in more than one organization. Several environmental NGOs of long standing are found in the Black Sea coastal city of Burgas, including SOS-Burgas, Ecoglasnost-Burgas, Burgas Ecology Foundation, Burgas Black Sea Club, and the Foundation Burgas-Ecology-Man, and they are largely led by the same few individuals. At one point, for example, two people—a former mayor of the municipality and the chief of the municipality's environmental department—had created or controlled virtually all the major environmental NGOs in the region (see, e.g., leader names listed in Penchovska et al. 1993, 1997). Though this intermixing of NGO memberships among activists is not necessarily of itself a problem, it does give a false impression of the total volume of environmental activism. It also illustrates the close connections that some nongovernmental organizations have to government bodies and also the creation of organizations to exploit available funds—topics to which I will return.

More important than the superficial focus on numbers is to understand these organizations at a more detailed level and in relation to both rural communities and the larger context in which NGOs find themselves. Specifically, what do these organizations look like, what do they do, and in what kind of political and economic environment do they operate? Bulgarian environmental NGOs range from small, local-level organizations concerned with place-specific issues to larger groups operating at regional or national levels and with more diverse sets of interests. Some environmental NGOs operate as single, independent organizations, while others consist of networks of branches or clubs in a particular region or throughout the country. Some organizations have become politically active, while others focus on less politicized concerns such as environmental education and nature conservation. As previously alluded to, some NGOs are associations of professional scientists or resemble environmental consulting firms. But others are groups of students wanting to work on environmental issues, or they may be local groups fighting against threats to their communities. Some NGOs have regular paid employees, while others are primarily volunteer groups, which may or may not occasionally hire people for particular projects. These groups are involved in wide-ranging activities and issues, including biodiversity conservation, sustainable agriculture, ecotourism, environmental education, pollution prevention, nuclear power, and management of water resources (Bulgarian NGOs 1995; Kodjabashev 1995; Penchovska n.d.; Penchovska et al. 1993, 1997). A 1995 survey of Bulgarian environmental NGOs found that nature protection and conservation was the most popular aim of these organizations, and 42 percent of projects reported in the survey dealt with natural objects and protected territories (Penchovska n.d.). My research focused on these organizations, although I also talked to people involved with numerous other groups to get a general sense of the Bulgarian environmental NGO community.[10]

The Politics of Environmental NGOs

The complicated relationship between environmental NGOs and political parties in Bulgaria deserves attention, because it potentially influences perceptions of the organizations by the general public and accordingly affects the relationships between the organizations and individuals or local communities. In theory, Bulgarian NGOs registered under the Law on Persons and the Family are not allowed to engage in political activities. This tends to be narrowly construed as a ban on having candidates for elected office (Kyutchukov 1995), however, and even here the situation can be confusing. As discussed earlier, environmental issues and environmental NGOs, particularly the organization Ecoglasnost, played a role in mobilizing opposition to the previous governing regime in Bulgaria. In the first round of free elections in 1990, several Ecoglasnost activists were elected to parliament from the Union of Democratic Forces (UDF) coalition, of which Ecoglasnost was a member. The connection between political parties and NGOs is thus particularly problematic in Ecoglasnost's case, but the general association can transfer to other organizations as well. Even though several environmental groups with "Ecoglasnost" in their name stress their independent, nonpolitical nature, this is not true for all. Koulov (1998:210), for example, describes the formation of two political parties with Ecoglasnost in their names in the Burgas area alone.

Since the collapse of the communist political monopoly, the anticommunist coalition has been less unified, and there have been several swings of political control between the socialists and the anticommunists and changes in the composition of ruling coalitions. Koulov (1998:206) suggests that the struggle between the UDF and the Bulgarian Socialist Party, and also the politicization of environmental decision making, particularly at the national level, has been detrimental for environmental efforts, making collaborative work at the local level difficult. Desai and Snavely (1998:40–41) similarly write that there was a loss of public regard for the environmental movement following public political battles among different factions, including a fight over whether they would sign Bulgaria's new constitution. A Green Party was formed during the early postsocialist period, subsequently dividing, and a socialist governing coalition in the mid-1990s included members of parliament aligned with National Movement Ecoglasnost (NM Ecoglasnost). According to one observer, NM Ecoglasnost representatives decided to participate in this coalition because they felt it would be more productive to try to work with the socialist government than to have no say at all in the ruling coalition. Other environmental activists who were initially involved in the political scene during the post-1989 period—as members of parliament, for example—have become disillusioned with the government and party politics and have refocused their activities on the nongovernmental sector. Friedberg and Zaimov (1998) point out that this means that many of those members of parliament in office when Bulgaria passed a strong environmental protection law and environmentally friendly constitution are no longer in place to advocate the preservation, enforcement, and enhancement of such legislation. Another incident that no doubt complicates this situation regards allegations that an Ecoglasnost leader, upon an investigation preliminary to his appointment to a government post, was an informer on Ecoglasnost for state security forces in the days leading up to the political changes (Daskalov 1998:15).

Some NGO leaders recognize the danger of being associated with party politics and specifically emphasize their organization's nonpolitical, nonpartisan nature, both in print and when talking to potential supporters or community members. Koulov (1998:209), for example, notes that the organization SOS-Burgas stresses its professional, politically neutral approach to environmental issues, and he writes that it has members from across the political spectrum. Similarly, the Bulgarian Society for the Conservation of the Rhodope Mountains describes itself in a 1994 newsletter as a "volunteer, nonpolitical organization," and the Bulgarian Association for Rural and Ecological Tourism presents itself as "nonprofit, nongovernmental, nonpolitical, non-ethnic, independent, and maximally open to everyone." This independence may be particularly important when groups work in rural areas, where support for the Bulgarian Socialist Party is often stronger, since environmental activities are often associated with Ecoglasnost's role in the fall of the socialist monopoly and formation of the anticommunist UDF.

Even while NGOs are proclaiming their independence, some people outside the NGO community continue to make connections between environmental organizations and political activities, due at least in part to the factors discussed above. This can count as a strike against the NGOs when a distrust of party politics is transferred to them as well. Alternatively, opinion about a particular party might be applied to an organization due to a real or presumed connection between the two bodies. For example, a leader of an environmental NGO that has received support from the Green Party said that this association has made some potential supporters reluctant to provide the group with financial resources. In another case, a Bulgarian acquaintance dismissed an environmental organization that I mentioned in a casual conversation by saying that it was a former communist-party youth group. This was not the impression I had from the participation of some organization members in the early 1997 protests against the socialist government, and even if it was the case, perhaps that was a forum through which some individuals managed to achieve environmental activity in the socialist era. Thus, politics continues to have the potential to affect the relations between NGOs and communities or donors.

Organization Size and Membership Characteristics

Many Bulgarian environmental NGOs have limited membership bases, and conditions in the country affect both the size of the organizations and the compositions of their memberships. First, NGOs are new on the scene. In addition, the general public's decision to join an NGO may be influenced by the extent to which they are seen as legitimate actors. Snavely and Desai (1995) suggest that gaining legitimacy is one of the greatest challenges facing the NGO community in Bulgaria, which also recalls the above discussion of NGOs and politics. Another possible influence is the severe economic crisis that has affected Bulgaria since 1989. Under these conditions, volunteerism may be hampered as people struggle for survival, and people are more likely to be actively involved under specific conditions.[11] A few people may be in comfortable financial circumstances that allow them to do so.

Some may be motivated to participate if their health or livelihood is directly affected by the issues being addressed by NGO activities—for example, mothers joining together to protest against hazardous air quality in Ruse in 1988, and community opposition to a water diversion in the Rila Mountains in 1994–1995 (Staddon 1998). Others may be able to generate funding for projects of interest by conducting the project under the auspices of an NGO. In the postsocialist aid environment, foreign donors are often more willing to fund nongovernmental organizations than governmental bodies, because NGOs are seen as democratic institutions that contribute to the development of civil society (see also Wedel 1998a). Government employees simply respond by creating NGOs. Students are another category of people often involved in Bulgarian environmental NGOs, and in some cases unemployed students or recent university graduates can gain needed experience and perhaps a bit of income by working with an NGO on an environmental project.

Indeed, because of the economic situation in Bulgaria, qualified experts and young people are sometimes involved in NGOs due to their inability to obtain professional research or teaching positions with government institutions or universities. These people have positive, useful activities in the country, and NGO involvement gives them an opportunity to gain experience in ecology or related fields. As Fisher (1996:58) more generally observes, employment with environmental organizations provides idealistic professionals in economically distressed countries with an alternative to dead-end government jobs or migration. In discussing NGOs generally in Russia, Bruno (1998:185) similarly writes that the people involved in these organizations are often from the academic intelligentsia and that this participation allows them to stay in their professional fields. One Bulgarian example comes from the Bulgarian Society for the Protection of Birds (BSPB). In 1997 this organization had five projects managed by young people. These students or recent university graduates wrote project proposals that were subsequently approved by the society's board. If a donor funded a project, its author (or authors) was then responsible for implementing and managing it. According to a BSPB official, along with the benefits of conservation achievements from the projects, the young people get hands-on experience with project management, thus giving the society experienced people to continue its work when the current leaders move on. In addition to explicitly recognizing the role played by such projects in giving young scientists practical experience during a period of economic distress, the official's comment is noteworthy for its concern about organizational sustainability.

Yet, one consequence of this phenomenon is that some prominent conservation and environmental organizations in Bulgaria can to some degree be characterized as professional groups or associations in which their members and leaders are often trained specialists—ornithologists, botanists, ecologists, foresters, and so on. For example, the Bulgarian Association for Ecological Agriculture—established in 1993 after a multidisciplinary conference on ecological agriculture—is largely composed of specialists from academic or research institutes around the country. Jancar-Webster (1998) similarly comments on the professionalization of environmental NGOs in Eastern Europe, and from a regional perspective, Bulgaria has fewer environmental grassroots organizations than the Eastern European average and a larger share of NGOs made up of professional environmentalists (REC 1997:16). While this professionalization may mirror the origins of Bulgaria's environmental movement, which largely started as a group of professional associations a century ago, it has potential consequences for relationships between NGOs and government institutions, on the one hand, and NGOs and rural communities, on the other. (Recall, too, that government employees sometimes create NGOs to increase their chances for funding of research projects or other activities.)

These organizations may have close ties to government ministries, universities, and research institutes because the NGO leaders were or still are employed by these entities. For example, the president of a prominent conservation NGO was also the director of the National Nature Protection Service for several years in the mid-1990s. Koulov (1998:211) similarly describes the intertwined relations and individuals involved in environmental management in the Burgas region, where three members of the Ecoglasnost Independent Union (an NGO with ties to the UDF) were elected to the city council and the NGO's chairman was appointed director of the municipality's environmental department.

These ties may give NGOs better informal access to the government and government policy or decision making than they might otherwise have, and in-house technical expertise and perhaps the language skills of the professionals benefit NGOs in both fundraising and conservation efforts. These government positions may also give NGOs more credibility with the general public, since the government is usually viewed as a responsible entity, while NGOs are seen as an unknown quantity or associated with politics (Koulov 1998). But it also means that NGOs may be less likely to oppose government initiatives (Jancar-Webster 1998:56), and Koulov suggests that this interdigitation of governmental and nongovernmental "counteracts one of the main objectives of the November 10, 1989, change in Bulgaria—the separation of party and state, political, and civil organizations" (1998:211). Such connections call into question Western assumptions about NGOs as part of civil society, in opposition to the government. But they may also be a fact of life for the Bulgarian NGO community, given the structure of foreign assistance and the lack of legitimacy sometimes accorded to the organizations based on their own merits.

Legal and Institutional Contexts for Bulgarian NGOs

In addition to membership size and composition, the extent to which NGOs can be embedded in society and the ways in which they carry out their activities are also affected by the institutional contexts in which they operate. This includes such things as whether they have the right to exist and their ability to participate in law drafting and decision-making processes. During the 1990s, the civil legal framework for Bulgarian NGOs was largely provided by the socialist-era Law on Persons and the Family (State Gazette No. 182, 1949). Since January 2001, they are governed by Bulgaria's postsocialist NGO law, the Law on Juridical Persons with Non-Profit Purposes (State Gazette No. 81, 3 October 2000). These laws address such issues as the creation, registration, membership, governance, and dissolution of the organizations as legal persons. Both laws specify two kinds of organizations, "foundations" and "associations," each of which have legal standing in the country following their required registration with the appropriate district court. A critical difference between these organizations is whether they have members—associations do, foundations do not. Different provisions in the law apply, depending upon how the organization is registered, the requirements for associations are more detailed. Representatives of two NGOs explained to me that their organizations were registered as foundations rather than associations because the process was easier and the status had fewer requirements. This shows how the law affects such a basic characteristic as whether to have members, although the director of one of these organizations was considering a status change that would allow it to have members.

The new NGO law further differentiates between associations established for the mutual benefit of their members and those established for public benefit. The kind of groups discussed here would likely be considered public-benefit organizations, and a number of provisions in the new NGO law specifically apply to them, in some ways reflecting problems identified with the earlier socialist-era legislation. For example, there are provisions for what is sometimes called "transparency" in "development-speak" (for example, the production and distribution of annual reports), and also for avoiding conflicts of interest (for example, members cannot participate in decision making that affects them or their relatives). A central registry of organizations identified as performing public-benefit activities has been established by the Ministry of Justice. These organizations are required to annually submit information to the registry, including an annual report, a certified financial report, and a list of officers. Yet, neither law is explicit about access to policymaking, nor do they reference such provisions in other laws. (The new law does contain references regarding financial matters, as discussed below.)

Other aspects of the institutional context in which Bulgarian NGOs operate that affect how the organizations carry out their activities and accomplish their objectives. Bulgaria's 1991 Constitution (specifically, articles 41, 43, 44, and 55) guarantees citizens and groups basic rights of public participation, including rights to free association, a healthy environment, information, and peaceful and unarmed assembly. The implications for the ability of NGOs to influence and participate in environmental decision making, particularly at the national level, are, however, less clear (see also UNDP 2001). Despite constitutional provisions about access to information, for example, some observers suggest that access to environmental data is still problematic (Kodjabashev 1995; Pickles and Mikhova 1998). Kodjabashev (1995:42) observes that problems with access to information could create a barrier to challenging the government over some activity. Quite simply, if a group could not obtain information about the existence of an environmental problem or environmentally damaging activity, it might not have a basis for a protest against the activity.

On top of this, no constitutional provisions or other laws give NGOs rights or guarantees to participate in legislative deliberation or the rule making of the parliament or other government bodies. For example, organizations cannot initiate legislation themselves; instead they must go through a member of parliament or the Council of Ministers. The latter ability often depends on the personal contacts of the NGO and its members, along with the willingness of a member of parliament to work with them, because access to policymakers—such as that seen in political lobbying by interest groups in the United States—is not at present institutionalized or considered a moral obligation (Kodjabashev 1995). One NGO activist commented to me that it was important to have friends in parliament, but even then it was easier to see them outside the National Assembly building, because of restrictions on who was allowed to enter the building and the need for an invitation to do so (see also Kodjabashev 1995:39).

When environmental organizations do participate in the development of legislation and other environmental policymaking, it sometimes happens more on an ad hoc basis— for example, when a particular person or group is invited to participate because of personal connections or reputed expertise on an issue—rather than through an institutional structure or institutionalized practice (Kodjabashev 1995). Bulgarian environmental NGOs participated in drafting the new protected-areas law, but some NGO representatives were critical of limitations placed on which organizations were invited to meetings. The Union of Private Forest Owners, established with the goal of participating in the development of a law to restore forests to their former owners, was involved in drafting that legislation. One may ask, however, whether the fact that the organization's president was a retired forestry director might have given him additional authority to speak on the issue. As Jancar-Webster (1998:87, my emphasis) points out in discussing Eastern European environmental NGOs generally, the professionalized groups have the possibility to be effective participants in environmental policymaking *provided that the infrastructure for doing so exists,* and in the Bulgarian case such structures and practices are not necessarily well developed.

The most complete public-participation procedures are included in the environmental impact assessment provisions of the 1991 Environmental Protection Act, containing the rights to information and to appeal administrative decisions (Friedberg and Zaimov 1998; Kodjabashev 1995). Their effectiveness was compromised, however, by a 1995 amendment to the law that exempts projects from this requirement when the vital interest of the population is at stake. Several NGO leaders identify this as a severe weakening of the law, and the organizations have made unsuccessful efforts to challenge the changes. Besides the length of time required for passage of new laws, environmental issues have fallen from the top of the government agenda as the economic crisis in the country has deepened. An NGO observer suggested that the parliament considers environmental legislation during lulls when it has nothing better to do. Similarly, Friedberg and Zaimov write that "in the last three years, political and economic difficulties have preempted any significant progress in further reform and enforcement of Bulgaria's environmental laws. With diminished resources, lack of social or political consensus, and the stalling of the movement toward civil society, the environment has largely fallen off the national agenda table" (1998:92). Such circumstances presumably have not made it any easier for NGOs to effect change through formal legislative channels.

There is also a general lack of precedent for NGOs to use the courts to accomplish environmental objectives and an incomplete set of legal tools for doing so (Kodjabashev 1995). Even if the Bulgarian legal system contains the basic authorizations for challenging activities that cause pollution or other environmental damage, their use is hampered by what Friedberg and Zaimov (1998:97–98) describe as the "lack of a culture of advocacy" in the Bulgarian legal arena. They suggest that progress in environmental matters in the country will require "a more crusading legal community" (Friedberg and Zaimov 1998:99), citing as an example the activities of several U.S. environmental NGOs and their lawyers in promoting environmental policy changes in that country. Beyond this, one sometimes hears general complaints about the inefficiency of the court system.

I came across one case in which a Bulgarian environmental NGO tried to use the courts to stop a project that it saw as environmentally detrimental. (Kodjabashev [1995] cites two other examples.) The Bulgarian Society for the Conservation of the Rhodope Mountains initiated court action against Bulgaria's Ministry of the Environment in 1993 in an effort to stop construction of a hydroelectric project on the Arda River, on the grounds that the ministry failed to comply with the environmental impact assessment provisions of the environmental protection law. After considerable delay, the Sofia municipal court judge issued a statement that she could not rule on the case, because the legal clause on which the society had based its case had not been in place at the time of the initial decision to allow the project. The Supreme Court upheld this original decision on appeal. (Although shelved at the time of the second decision, the project has since been revived and construction has started.) The outcome of this court case was less than definitive and thus may not provide the precedent that some environmental activists had hoped a victory for the environment might have set.

Kodjabashev (1995) and Koulov (1998) suggest that NGO access to and participation in environmental decision making at the local level is somewhat better than that at the national level—provided that the political polarization mentioned earlier does not get in the way. For example, memos of understanding that have been signed between some local government administrations and the Bulgarian Society for the Protection of Birds, agreeing that they would work together either generally or on specific issues such as wetland restoration or environmental monitoring. As well, municipalities that are short of funds may work with or even create environmental or local development NGOs as a way of obtaining money for projects, since it is often easier for nongovernmental groups to obtain grants than it is for government bodies to do so. For example, the Burgas municipality established the Foundation Burgas-Ecology-Man to help the city solve its ecological problems (Koulov 1998).

Livelihood Strategies of Bulgarian Environmental NGOs

Another critical aspect of the institutional context in which Bulgarian NGOs operate is the availability of financial resources for their work. This can influence the form and activities of NGOs and the extent to which they are oriented to the concerns and interests of rural communities versus donor organizations. As Price (1994) points out, even those NGOs that rely primarily on volunteers need to purchase materials, perhaps pay for office space, and finance projects, and thus they need to maintain a steady flow of resources. Securing funding can consequently be a critical and time-consuming activity for any NGO. Bulgarian environmental organizations are no exception, and they operate in a context in which it is difficult to obtain funding from in-country sources, in part due to the ongoing fiscal crisis and the absence of a tradition of philanthropy in the country over the last half-century. The ability of NGOs to generate funding within the country may also be influenced by the extent to which the general public sees them as legitimate and effective actors.

There are both practical and legal constraints on raising funds from sources within the country. One common source of funds for environmental organizations in the United States, for example, is membership dues. The Bulgarian Society for the Protection of Birds charges membership dues, despite the fact that it costs them more to maintain these members (e.g., for collecting the dues and providing membership benefits such as the organization's newsletter) than the society generates through the dues. Organization officials believe it is nonetheless important for the members to show support for the society in this fashion. The leader of another organization reported that it initially collected dues, but stopped when the administrative costs exceeded the income generated from their dues-collection efforts. It may start collecting dues again when its membership base reaches a size that would make this a moneymaking enterprise. The dues collected by a third organization are retained by its local branches, which operate on a volunteer basis, and play a role in funding local activities. For Bulgarians, dues are typically low, with the amounts sometimes based on a specified percentage of minimum monthly-wage levels. Foreign members pay more, and thus they may be more lucrative sources on an individual basis, but their numbers are small. Thus, the small size of many Bulgarian NGOs, combined with the country's financial situation, makes the collection of dues impractical as a major fundraising strategy for most organizations.

The socialist-era Law on Persons and the Family, and its interpretation by district courts, also created barriers to fundraising during the 1990s. Early in my fieldwork several NGO leaders complained that they were not allowed to engage in activities that might be construed as doing business and that this restriction limited their abilities to generate funds for the groups' activities (see also Kyutchukov 1995). This included, for example, some things done by environmental NGOs in the United States and Western Europe to raise funds and promote their organizations and activities, such as selling T-shirts, books, or posters and running nature tours. At least one Bulgarian group got around this prohibition by creating a parallel trading company that then made donations to the NGO. Other organizations have considered working with an existing business or organization that has the right to engage in such activity and then having the business or organization donate a portion of the proceeds to the NGO. With passage of the new NGO law, some of these financial issues have been clarified for public-benefit organizations. The law states that these organizations can engage in economic activities if they are related to the main activity for which the NGOs are registered and if the revenues are used for achieving the registered purposes. It further declares that the government may establish tax, credit, customs, interest rate, and other benefits for these organizations, but the specifics of those benefits will be spelled out in other laws.

Foreign-Donor Support for Bulgarian Environmental Organizations

Although some Bulgarian organizations obtain limited financial support through fundraising projects, membership dues, donations from Bulgarian companies or individuals, or from local or national government ecofunds, most Bulgarian conservation NGOs rely largely on foreign funding to support their activities.[12] Often this funding comes from the Regional Environmental Center for Central and Eastern Europe (REC), or from international, Western European, or North American NGOs or governments. Indeed, the REC is frequently mentioned as a source of funding for environmental NGO activities in Bulgaria. The REC was established in 1990 as an independent nongovernmental, nonprofit foundation by Hungary, the United States, and the Commission of the European Communities, and additional countries have joined as sponsors since this time. Its initial mission, which reflects Western concern about the development of democracy and civil society in the region, was "to promote cooperation among diverse environmental interest groups in Central and Eastern Europe; to act as a catalyst for developing solutions to environmental problems in this region; and to promote the development of civil society" (REC 1993). The local grants program, administered out of country program offices such as one in Sofia, provides small sums of money in local currencies to organizations for institutional development (e.g., operating expenses, registration, equipment) and grassroots projects, while the earmarked grants program, run out of REC's main office in Hungary, provides larger grants for projects on themes such as nature conservation or pollution prevention. Another earmarked program specifically encourages cooperative environmental projects involving NGOs in at least two Eastern European countries.

Another popular funding source during my 1996–1998 fieldwork was the USAID-sponsored Democracy Network Program. This is an example of both the ideology of Western aid to Eastern Europe and the way in which funding availability can drive the type of project undertaken. The purpose of the Democracy Network Program, which is being implemented in ten postsocialist countries, is to assist local NGOs with funding, training, and technical assistance. In its first phase, the program had $1.5 million to distribute to Bulgarian organizations over a three-year period. All funded projects must have a democracy-building component, which is part of an ideology about NGOs, civil society, and democracy being imported to Eastern Europe by Western aid agencies. The first round of funding focused on democracy-building NGOs—those involved in human rights or voter awareness, for example—and the second round included NGOs with social safety-net programs as well. The third and largest round awarded in fall 1997, which was also the last round in the program's first phase, included environmental organizations, and several environmental groups received funding. These funds went to established and young organizations, both inside and outside of Sofia. Some organizations received support for institutional development, a significant concern for many environmental NGOs in Bulgaria, and others received funding for specific projects.

While this funding source is neither nongovernmental nor explicitly environmental in nature, it is mentioned for several reasons. During spring 1997, it seemed that every environmental NGO representative I spoke with was preparing a proposal to send to the program, which is to say that it attracted considerable attention from the organizations. It is also an example of a program that requires specific kinds of activities from the NGOs it funds—in this case, democracy building. In some cases the connection is obvious, like a project to improve public participation in environmental decision making and policymaking. In others, the connection is less direct, such as how to turn a project to promote ecotourism into an exercise in democracy building. Like the REC but unlike many other external donors to Bulgarian NGOs, the Democracy Network Program has an in-country presence, and NGO officials are able to speak to the project officers in Bulgarian, since all of the project officers, trainers, and technical assistance specialists are Bulgarians. This increases the program's accessibility to groups with limited foreign language skills. Once funded through the program, organizations remain in the network and have access to its training programs and technical assistance resources, so that knowledge is transferred as well as money. Most NGO representatives I talked to who have been funded by this program have been pleased with the resources it offers them, suggesting that this program has done a good job of relating to its constituents.

Although a few grant-giving organizations and programs such as the REC and the Democracy Network provide Bulgarian environmental NGOs with ongoing operating expenses or help with institutional development, much of the external funding available is for special projects. For instance,, the Dutch Foundation for Education and Training provided financial support to the Sofia-based Foundation for Ecological Education and Training for an ecotourism seminar. The topic for this seminar, which was a major activity for the Bulgarian organization in 1996, was suggested by the foreign donor to the foundation, which had no previous experience with ecotourism. This illustrates the potential for international funding to influence NGO activities. Similarly, the activities of the Bulgarian Association for Ecological Agriculture have been largely restricted to holding conferences, because their funding applications for other types of projects have been unsuccessful.

Another example of funding availability driving organizational activity in an unexpected direction comes from the University Rescue Squad. This organization of current and former students started its postsocialist life using technical mountaineering skills for activities such as mountain rescue and responding to contamination incidents. The organization's president described it as being in a constant state of financial collapse, not knowing, for example, how it would pay the telephone bill. Eventually it came into contact with the Bulgarian Red Cross, which started to provide financial support for projects, specifically ones that involved feeding homeless children in Sofia and working with orphanages in the country. The president said that organization activists had prepared "100 proposals" for projects, but the only ones funded were for projects working with orphans and homeless children. He has nothing against such activities, but sees them as a poor use of the technical skills of the organization's members. Bulgarian NGOs, like rural residents, are operating in an uncertain environment and must develop their own postsocialist survival strategies. This can include undertaking projects for which funding is available, even if they are not necessarily in areas where the organizations have specific expertise.

Reliance on project-oriented funding creates the potential for additional problems for the NGOs (see also Sampson [1996] on the social life of projects in Albania). The project nature of funding can result in uneven levels of activity as organizations move from project to project, punctuated by periods of grant writing. There is some danger that projects may be written to respond to the requirements and interests of donor organizations, as illustrated by earlier examples. The need to obtain external funding may lead NGOs to focus on external concerns at the expense of local interests. The organizations that survive are sometimes the ones that are good at writing proposals and have learned the necessary project jargon—an area in which the professional nature of some groups may be an advantage. The organizations are not unaware of the issue of "project-speak," as was quite clear in a conversation I had one day in the office of a small but successful conservation organization. While discussing the English-language skills of an activist and whether they were improving, one of his colleagues cynically commented that perhaps it was just his facility with the terminology needed to write successful grant proposals. Another potential consequence of this reliance on external funding is that competition for limited funds may lead to competition among organizations rather than cooperation (Kodjabashev 1995:46).

The fact that most Bulgarian environmental organizations rely on external funding has consequences for the structure of the NGO community as well. Specifically, new NGOs are sometimes created and specifically tailored to maximizing access to external funding resources, resulting in an artificial increase in numbers. One example of what might be called "niche marketing" by NGOs is the number of organizations in Burgas that involve the same two individuals. To the outside observer, it appears that different organizations were created in response to different fundraising opportunities, in other words, to fit into different financial niches. Similarly, the Sofia branch of the Green Balkans Federation registered as a separate organization to maximize its access to funds. The idea was that as two separate NGOs, each organization could apply to the same funder, thus expanding funding possibilities; although later they were more concerned about working together. The youth section of the Bulgarian Society for the Conservation of the Rhodope Mountains similarly split off as a separate organization to avoid having two units of the same organization applying to the same funder. Other reasons cited for this division by the new organization's president were the relocation of the society's headquarters from Sofia to a town in the Rhodope Mountains, which made activity coordination more difficult, and age differences between individuals in the youth section and the larger society. Such examples illustrate how funding availability influences the number of organizations.

Environmental NGOs and the 1996–1997 Financial Crisis

Further insight into Bulgaria's environmental NGO community is gained by exploring what happened to some organizations when the country's economic condition worsened in 1996 and then nose-dived in February 1997.[13] For some NGOs, working in Bulgaria in 1996–1997 also included dealing with the political and economic context, in particular changes of government and the activities of groups that Wedel (1998b) calls "cliques" and are locally referred to as "the mafia." Particularly telling is how different organizations were affected in different ways. Indeed, one interesting, although not unexpected, response of some organizations to my questioning about this situation was that they had not been significantly affected by the currency crisis, followed by details about how they had been otherwise impacted by working in Bulgaria at the time. This was particularly the case for established Sofia organizations with strong international ties, as the examples below illustrate.

Borrowed Nature is a small Sofia-based organization that was established in 1992 and has strong international ties. It focuses on raising public awareness about environmental issues, involving citizens in environmental decision making, publishing materials on environmental topics, and participating in national and international events. Although less directly involved in biodiversity conservation than other groups discussed here, Borrowed Nature has been involved in activities to support environmental NGOs in Bulgaria generally, and it also advocates NGO participation in decision making. Borrowed Nature and its members are involved in the United Nations Development Program's Capacity 21 Program on sustainable development, and one of its first projects promoted bicycling as an alternative form of transportation in Sofia. In 2001 its portfolio also included a project related to biodiversity conservation in the Rhodope.

In responding to my questions about the influence of the 1996–1997 political and economic situation, the activists I interviewed first expressed their concerns about personnel changes at several ministries as a result of the change in government. There had been a strong candidate for environmental minister backed by the NGOs, they said, but someone else was eventually appointed.

They also wondered who would be appointed to head the Ministry of Education, the Ministry of Territorial Development and Construction, and the National Electrical Company. After these appointments, they expected additional personnel changes at lower levels in these agencies. Borrowed Nature was also affected in a minor but nonetheless inconvenient fashion by the weaknesses of Bulgaria's banking system. A foreign partner sent them some money by bank transfer, but their Bulgarian bank had trouble receiving it, and the funds and associated project were consequently held up for 40 days. In another case Borrowed Nature was working with a Dutch partner. It seemed to the Bulgarians that the most efficient and safest way to receive funds from this partner was to set up an account with a Dutch bank that had a branch in Sofia. However, that business-focused bank said that Borrowed Nature was not a large enough customer to open an account. So, even if the banking system's condition did not cause the NGO to lose money, it was a source of frustration.

The **Wilderness Fund** is a small but established Sofia-based NGO that works to preserve the uniqueness of Bulgaria's natural world by strengthening the protected-areas network and ensuring long-term protection of valuable species and their habitats. It was established in October 1989 and officially registered in early 1990. As in the preceding case, foreign ties helped to protect the organization from the financial crisis. It receives much of its funding in hard currency, which it placed in what one activist described as "the only safe bank"—the Bulgarian foreign trade bank. Most of the organization's leaders have also been involved long enough that they have become used to working in the Bulgarian context, although the situation proved more difficult for their younger and less experienced project managers. The case described below also illustrates how foreign connections and reliance on foreign funds can periodically cause headaches for the organizations that depend upon them.

I visited the offices of the Wilderness Fund on a day when a couple of officials, one of them a less experienced project manager, were in the midst of a disagreement with a funder. The conflict was over an international project for the protection of the Balkan bear population, which involved NGOs in Albania, Macedonia, and Greece as well as two Bulgarian organizations. The Wilderness Fund was the lead organization. Due to conditions in

Albania at that time—the collapse of pyramid investment schemes and the country's subsequent descent into anarchy—the organization wanted to reallocate a moderate sum of money within the reasonably large project budget. The strict budgeting rules of the donor made this difficult, however, and the NGO activists involved suggested that more flexibility was needed for working in the Balkans. Beyond financial headaches, however, most activists and officials from this organization expressed concern about the new government—the extent to which the environment would be a priority and the kinds of environmental issues on which it would focus. Thus, the Wilderness Fund was less affected by the economic crisis in Bulgaria and more concerned about what the associated political changes would mean for the environment.

The Sofia-based **Bulgarian Society for the Protection of Birds** (BSPB) has branches in several localities around the country, including in the Rhodope, and a substantial base of paid members. It has strong ties to the international bird-protection community, which provides both financial and technical assistance. The work of this organization, the first environmental NGO to officially register in 1988, focuses on biodiversity and nature protection. Although its primary focus has traditionally been bird protection, such work also includes the protection of bird habitats, and the organization has more recently diversified its activities to attract a broader range of people. A BSPB official reported that the organization was not significantly affected by the 1997 economic crisis, adding that its strong Western contacts helped. Indeed, the influence of regular foreign funding on the financial situation of this organization, as well as Borrowed Nature and the Wilderness Fund, is also seen in their 1996–1997 moves into new offices that were nicer, more centrally located or both. In the case of the bird protection group, it now owns the office—a ground-floor apartment on a main road relatively near the city center—rather than just renting office space.

The BSPB has, however, been affected by other aspects of the situation in postsocialist Bulgaria. The society reportedly received phone calls from rackets offering to protect its previous office. Another problem concerned telephone services in both its old and new offices. The group wanted a second phone line in its new office, but given the condition of Bulgaria's telecommunications network, that would require either a long wait or paying a bribe to get the line more quickly. For the old office, the organization had paid $400 to get a phone line installed because the waiting list took forever. The money is less of an issue than the principle of the matter, said one leader. (Contrast this with the University Rescue Squad, which is in a constant state of financial collapse and rarely knows how it will pay its phone bill.) A similar dilemma was encountered in choosing an insurance company for the organization's vehicles. The question was, should the vehicles be insured with the state company, which was seen as less corrupt but might not pay on a claim quickly (if at all), or should they choose one of the large private insurance companies, often described as being controlled by mafia groups? Following this discussion about telephones and insurance, the activist I spoke with said that he looked forward to the day when life in Bulgaria was more normal—when Bulgarians did not have to deal with these concerns and had more time to pay attention to environmental issues.

The **Green Balkans Federation** is a Plovdiv-based NGO that focuses on biodiversity conservation in Bulgaria and on the Balkan Peninsula, including involving the public in solving nature-conservation problems. Besides the main office in Plovdiv, the Green Balkans Federation has offices in Burgas, Stara Zagora, and Sofia. Although well established, it was more affected by the situation in 1996–1997 than the three previously discussed organizations. In responding to my question, the organization's president reminded me that, first, the Green Balkans activists were affected themselves, as they lived in Bulgaria (Bruno [1998:185–186] also makes a similar point). Beyond that, the organization lost some hard currency deposits when First Private Bank failed, and it lost additional money due to rapidly changing exchange rates. Another way the organization was affected by the postsocialist situation is that it had had to move its office. In contrast to the moves of the three organizations discussed above, this was not an upgrade to a nicer, larger, or more centrally located facility. Rather, the precollectivization owner of the building in which the organization formerly had an office wanted it back, and so Green Balkans had to move. The organization's leaders were concerned that this move and the associated address change would disrupt some of their foreign contacts, again emphasizing the importance of external relations.

The **Bulgarian Society for the Conservation of the Rhodope Mountains** (BSCRM) was established in 1990 with the goal of protecting the natural and cultural heritage of the Rhodope Mountains. Besides Sofia and Plovdiv, the society has branches in villages and towns around the Rhodope. In part due to concerns about rental fees, the society moved its office in the mid-1990s from the Institute of Ecology in Sofia, where its president worked, to the president's home in the central Rhodope town of Chepelare. Initially more active, the society had slowed down somewhat by 1997 from its early days, particularly following the president's death in June of that year. Like Green Balkans, the economic crisis did not cripple the organization, but it was affected more than the first three cases described here. The society lost some money in a bank failure and delayed a regular membership meeting scheduled for spring 1997 because of the costs and logistics of getting representatives of the local branches (scattered around the Rhodope) to a central meeting location during the chaos. At a rescheduled general membership meeting in late 1997, a founding member expressed concern about the small number of attendees in comparison to the number of people and the enthusiasm at earlier meetings. The organization may be seeing the impact of postsocialist economic conditions, as the ordinary people who make up the membership of BSCRM's local branches are now otherwise busy with economic survival. The economic and political situation also has affected the society's involvement in ecotourism, although it is difficult to sort out how much of this is a result of the economic situation and how much is due to the death of its president, who was a major force behind the organization.[14]

The five organizations described above are comparatively large, active, and established. Smaller NGOs often had different kinds of troubles—perhaps reflecting their small size, inexperience with organizational management, and limited financial resources. The activists in one young organization that promotes rural and ecological tourism told me at one point in 1997 that they would not be traveling as they had planned, because their travel money had been spent on computer paper. Another new organization had trouble finding space to rent for an office with the funds allotted in their institutional-development grant from the local REC office because of increased apartment rental prices in Sofia. Other environmental NGOs undoubtedly lost money in the collapse of the banking system, as happened to a couple of the organizations described above. And even if an organization did not lose money in a bank failure, if its funding was from a source that made grants or gave donations in local currency—like the local REC office—or if money had been exchanged for local currency before depositing it in the bank, the funds could have lost much of their value with the currency devaluation and near hyperinflation. The resulting financial situation may have also caused other organizations to lose or change the location of their offices because they could no longer afford them, and this may have affected the ability of funders and potential members to contact them.

The most obvious dichotomy to emerge from this discussion of the ways in which Bulgarian environmental NGOs were affected by the situation in the country in 1996–1997 is seen by comparing the impacts on small organizations relying on domestic sources of funding with those on the larger, more established organizations having greater reliance on external sources. The latter organizations were less affected financially by the crisis and more concerned about the political situation. This is not to imply that the other organizations were not concerned about politics; it is just to observe that the economic conditions were less important for the more financially secure groups and they could worry about less immediate issues. Although similar financial issues (i.e., making bank transfers) may have confronted smaller organizations, the issues were more complex and varied for these larger and more established NGOs: how to make international bank transfers, how changes in government personnel would affect environmental policy and policymaking, and what to do about car insurance or telephone service.

Discussion

Bulgaria's long history of nature conservation activities and the involvement of nongovernmental organizations in environmental concerns dates back nearly one hundred years, although the focus of these efforts and the involvement of independent actors therein ebbed and flowed over the century. With the creation of the National Nature Protection Service, passage of a postsocialist law on protected areas, development of park management plans under the auspices of externally funded projects, and establishment of management authorities for some parks, one can see the beginnings of an institutional structure for conservation and a system of protected areas in the postsocialist period. What this means in practice and on the ground is less clear, however. Responsibility for protected-area management until recently was decentralized. Most park management plans are new and thus untested, and people living in and around these parks and reserves may not know about their existence or the land-use regulations associated with them (Mihova 1998:714). Although progress is being made toward institutional strengthening of the protected-areas system with the help of international sponsorship of projects such as the GEF Project and the Bulgarian-Swiss Biodiversity Conservation Program, sustainability will be an issue once the foreign donors pull out. Here it is telling that both of these projects were extended beyond their initially anticipated duration.

Like conservation activities generally, NGO involvement in environmental issues in Bulgaria dates back to the establishment of the first parks and reserves, although this long history is rarely given more than passing notice in discussions of postsocialist environmental politics in Bulgaria. While independent action on environmental issues was difficult during the socialist period, developments during that era do affect the shape of today's NGO community through the training of environmental professionals and the participation of individuals in state-sanctioned organizations, some of which declared their independence once such action was possible. Indeed, in some cases, NGOs have been created in response to the lack of action by these government authorities or their environmentally threatening policies.

With the fall of the communist party's political monopoly, conditions for the reemergence of an independent environmental sector liberalized significantly. Even so, the larger context continues to affect what this community looks like and its ability to accomplish its objectives. The ability of Bulgarian NGOs to participate formally in environmental decision making is limited, for example. (In discussing conservation in Madagascar, Gezon [1997] similarly notes that the larger institutional context for conservation can influence the ability of projects to achieve their objectives.) The NGO community in Bulgaria and the kinds of projects undertaken by the NGOs are also influenced by the financial context in which they operate. This has several consequences. Many organizations are small, particularly in terms of the number of active members, and their often professional memberships do not necessarily mirror the composition of society at large. Most rely largely on project funding, so that donor priorities may influence the activities undertaken. The number of environmental NGOs in the country is at least partially a function of the availability of funding for their work, and competition between organizations for limited funding may sometimes lead to secrecy about their activities and a lack of desire for cooperation (Kodjabashev 1995:46; Bruno 1998:185 makes a similar observation based on Russian material). An examination of the ways in which different organizations were specifically affected by the 1996–1997 political and economic crisis highlights variation within the community along with the constraints imposed by local conditions. Most significantly, it shows the way in which secure external funding buffered some organizations from the immediate financial crisis, but also that these organizations were still affected by other aspects of postsocialist conditions. Despite limitations on their ability to directly influence environmental policy in the country, NGOs are occasionally involved in policymaking with the potential to eventually influence resource use and management on the ground, including in the Rhodope. These include, for example, participation in crafting the biodiversity strategy, the protected-areas law, and the forest restitution law. Such policymaking participation also includes drafting proposals for new protected areas. The organizations also are conducting a variety of projects in different places in the country, and chapter 5 takes up these more concrete activities, particularly in the Rhodope, and the connections among and between the donors, organizations, and local communities involved. Before doing so, however, the next three chapters turn to the story of one Rhodope village in order to provide a picture of natural resource use in one of the rural landscapes that NGOs seek to protect.

2

Landscape, Community, and Economic History in the Central Rhodope

To reach the central Rhodope Mountains and the village of Zaburdo, the main route from the Bulgarian capital of Sofia is via the towns of Plovdiv and Asenovgrad. The highway between Sofia and Plovdiv is a good one—recently resurfaced, with two lanes in each direction. Traffic is rarely heavy, although shiny black Mercedes, racing past at what seem to be the speed of light, draw my attention. Once out of the upland basin in which the capital sits, the highway passes through flat agricultural land with mountains on both sides. Just before Plovdiv, a road sign points south to the Pamporovo ski resort and the regional center of Smolyan. The road gets narrower after I make this turn—becoming a two-lane, undivided one—but the terrain does not change much until I reach Asenovgrad. On the outskirts of town, I pass a large complex for processing nonferrous metals, set in a field of bright yellow sunflowers, and I note with some amusement a sign for unleaded gasoline that sits beneath the lead smelter's smokestacks. If I'm driving my blue Russian station wagon, I often stop at the Shell station for a fill up of leaded gas and my favorite Greek ice-cream bar. Alternatively I could purchase a bottle of Mavrud wine, for which the town's winery is famous, at the station's modern minimart. Bridal shops seem to be on every corner in this market town, which also has the last traffic light for a long way.

Continuing south from Asenovgrad, the road abruptly enters the Rhodope Mountains, following the winding but gradual path of the

Chepelarska River. In the short distance between Asenovgrad and the village of Bachkovo, the valley is so narrow that the road passes through two tunnels blasted into the hillside; one is long and unlit, which makes me nervous with my Lada's weak headlights. Just outside Bachkovo, a roadside sign identifies the Chervenata Stena Biosphere Reserve, one of 17 such UNESCO-recognized sites in Bulgaria, and the road then passes Bachkovo Monastery, the country's second largest. Nine kilometers farther is the junction leading to the mining town of Luki, the Kormisosh hunting reserve where former communist leader Todor Zhivkov shot a world-record-setting wild boar, and the village of Manastir. The last is famous for being the highest village in Bulgaria and the ancestral home of its recent president.

Continuing along the main road, I pass a few settlements and other signs of humans—small orchards, farms tucked into pockets of flat land, abandoned fish farms, a resort community known for its mineral springs, roadside inns. But the most obvious thing is the natural environment—the rushing river, rock outcroppings, narrow valleys, deep blue sky, and steep, tree-covered hillsides. The dominant color is dark green from the thick forests that dominate the region, but in the spring some slopes are dotted with purple lilac blossoms, and in the autumn the reds and yellows of changing leaves on deciduous trees brighten the landscape. As I turn another corner the valley widens, the fields and orchards of the villages of Pavelsko and Hvoina unfold, and then the valley narrows again.

About ten minutes south of Hvoina, an inconspicuous turn to the right is marked with a small sign that says Zaburdo. I make the turn, and the road gets steeper and narrower. This road, too, follows a small, winding river and at times is cut into the steep hillside. There are many twists and turns, but little sign of people. After ten kilometers, the road forks. The right fork leads to three natural stone bridges designated as a natural landmark and a large tourist chalet, but our route is to the left, going around the large bald rock that dominates the landscape. After this point I might encounter a flock of sheep, a herd of cows, or a lone man and his mule transporting hay or firewood; but then again I might not meet anyone. Three kilometers after the junction, there is a former trout farm that has been turned into a modern villa complete with swimming pool by an outside businessman, followed by the ruins of a long-abandoned grain mill, then a small barn and bean patch, and finally another sign saying Zaburdo, this one larger than the last.

The journey from Sofia has taken just under four hours, although the village is still not in sight. As I turn yet another corner, it finally comes into view. The street is lined with wooden barns and multistoried stone or brick houses, with roofs of red ceramic tiles or lichen-covered stone slabs. The houses are packed closely together with only the smallest of yards, due to the steep, rocky slopes of the narrow valley. Many houses have balconies, and from some of them hang freshly washed clothes or hand-woven blankets and rugs. The street is lined with stacks of firewood, Lada cars, large Russian farm trucks, and wooden horse-drawn carts, and it is best to watch your step to avoid landing in the cow manure left behind by the neighborhood herd. On sunny evenings small groups of people sit on benches outside their houses, periodically changing locations to follow the last rays of the sun while they chat, perhaps knit, and wait for the cows to come home. Some children dressed in T-shirts and jeans or sweatclothes are playing volleyball in the street; their improvised net is a rope tied between the drainpipe of a house and the side-view mirror of a large Russian truck parked on the other side of the street.

This travelogue provides a sense of the territory and distance crossed when traveling from Bulgaria's urban capital, Sofia— with its noise, air pollution, bustling crowds, NGO offices, and modern stores—to the rural mountain village set among thick forests, along a rushing stream that serves as our window into life in the region. The roads are good, and under normal conditions the journey is not arduous, but the distance is clear. Even for the average Bulgarian, much less a foreigner, the Rhodope can be something special and also somehow unknown. The folk music is famous from this one-time home of the mythical Orpheus, and the first Bulgarian song in space, sent on the spaceship *Voyager* as an example of Earth culture, is from these mountains. Aspects of material culture are also notable, including hand-woven textiles, brightly colored folk costumes, or the traditional architecture preserved in the ethnographic village of Shiroka Luka. The region is also known for its recreation opportunities, which include hiking along mountain trails in the summer, visiting mineral-spring spas for a relaxing cure, or skiing at the Pamporovo resort complex in the winter. Local people ski as well, and it was a happy moment for many Bulgarians when a young woman from the central

Rhodope town of Chepelare won a gold medal in the biathlon in the 1998 Winter Olympics in Japan—Bulgaria's first-ever gold medal in a winter Olympics contest.

But beyond that, the picture—particularly of village life—becomes less clear for those who have not spent time in the region. Bulgarian literature, history, and film portray this as a land of hard-working shepherds and small farmers, some of whom suffered innumerable injustices at the hands of bandits and "Turkish oppressors" during five centuries of Ottoman rule (e.g., Donchev 1997; see Iordanova 1998). While details of that experience are unclear and likely more complex than this simplistic discourse of the "Turkish yoke" (or slavery) implies, it is true that rural and sometimes remote villages in the Rhodope are today home to a large share of Bulgaria's Muslim population, either the descendants of Turks who settled in the region or Bulgarians who converted to Islam during Ottoman times.[1] Yet, in other ways the region's residents are not so different from anyone else. Literacy rates are relatively high; evenings may be spent watching television or reading the newspaper; children may dream of growing up to be sports stars, doctors, or policemen; and parents worry about their children's educations, future professions, and the possibilities for marriage and grandchildren.

The story of one rural Rhodope community, told through in-depth ethnography, provides insight into rural life and the use of natural resources that conservation efforts aim to protect. After describing the physical setting and environmental context, this chapter focuses on the village of Zaburdo.[2] It briefly introduces the village and its residents before turning to history, in particular to the history of the village economy and natural resource use. This discussion moves between events at local and national levels in demonstrating the ways in which village life has been affected by external forces, particularly in the century prior to 1989, and how local residents have more or less successfully responded to these changes.[3]

For many centuries, Rhodope villagers have practiced mixed subsistence strategies in their mountainous environment. Agriculture, pastoralism, and more recently forestry have provided the foundations of the economic system. Over time, however, ownership of resources and the organization of production have been influenced by political and economic factors external to the village, and new opportunities for nonagricultural employment have arisen. Local residents gained private ownership of agricultural lands and some forests in the mid- to late 19th century, that is, toward the end of Ottoman rule, and then they lost this control after World War II and the rise to power of the communists. These changes in regimes of control, along with changing economic opportunities, in turn influenced the ways these resources were exploited, and also in a sense affected the identities of villagers. Zaburdo residents changed from private smallholders to wage workers on the cooperative farm or in nonagricultural enterprises, and they were also increasingly incorporated into the larger cash economy over the course of the 20th century. At various points in time, externally imposed factors such as religious-conversion campaigns and establishment of the cooperative farm have led to outmigration from Zaburdo, which in turn has influenced the community's demographic composition.

This historical background is also important for recognizing continuities and changes regarding the postsocialist situation in terms of natural resource use and livelihood strategies, to be discussed in subsequent chapters. Although few villages can realistically be described as typical, my experiences traveling elsewhere in the region suggest that many mountain communities in the Rhodope are broadly similar in physical infrastructure, social organization, and economy. If anything, residents of Zaburdo are more dependent on use of the local landscape than residents of other villages, due to a lack of employment alternatives. But for this book's broader goal of understanding the village economy and strategies of resource use and how they relate to biodiversity concerns, this is the kind of village that might be most affected, and thus whose resources use it is most important to understand.

The Landscape of Orpheus

The Rhodope Mountains occupy a broad area extending across Bulgaria's southern border into Greece, measuring 240 kilometers from east to west and as much as 100 kilometers from north to south at their widest point (see Fig. 1). The largest in area of Bulgaria's mountains, the Rhodope cover an area of 14,735 square kilometers in the country. (About 80 percent of the range—which

Figure 1. Map of Bulgaria (prepared by University of Kentucky Cartography Lab)

covers a total territory of roughly 18,000 square kilometers—is in Bulgaria; the remainder is in Greece.) Compared to other mountain ranges in the country, which include some of the highest peaks on the Balkan Peninsula, the Rhodope are relatively moderate in elevation. They average 785 meters above sea level overall, and about 1,100 meters above sea level in the western section—in which Zaburdo is located. The high point at Peak Golyam Perelik in the south central part of the range reaches a modest elevation of 2,191 meters (Danchev 1998; Ministerstvo na Teritorialnoto Razvitie i Stroitelstvoto 1994:41; Perry 1995:115).

Consequently, the mountains lack the alpine character seen in ranges that experienced substantial glaciation. Instead, one encounters a typical karst landscape, with caves and deep gorges in the western part of the range. The eastern section is composed primarily of younger igneous rocks, and the more significant landscape features are rock mushrooms and pinnacles weathered by wind and rain (Ministerstvo na Teritorialnoto Razvitie i Stroitelstvoto 1994:41; Perry 1995:115–116). Much of the central part of the Rhodope consists of a series of ridges separated by valleys, running along narrow rushing watercourses. Some of the ridges trend north–south, others east–west. To the west are extensive highland plateaus. Natural lakes are rare in this relatively well-watered region, although one occasionally encounters large reservoirs associated with hydroelectric power generation.

The moderate elevation and extensive territory also have implications for settlement patterns and human use of the landscape. Cultivation can potentially take place on the highest ridgetops, and for pasturing livestock, distance or topographic features such as cliffs are more important constraints than elevation and associated vegetation changes. In contrast to other ranges in the country, where settlements typically ring more compact or linear upland areas, much of the Rhodope is dotted with hamlets, villages, and small towns. Rather than alpinists or rock climbers, these mountains attract hill walkers, cavers, skiers, and bird-watchers from near and far.

The Rhodope Mountains are also home to numerous and diverse animal and plant species due to the large area covered and the variety of terrain and climate.[4] An official from Zaburdo's hunting society, for example, provided the following list of locally occurring wildlife: rabbit (*Lepus europaeus*), roe deer (*Capreolus capreolus*), wild boar (*Sus scrofa*), red deer (*Cervus elaphus*), wild goat or chamois (*Rupicapra rupicapra balcanica*), brown bear (*Ursus arctos*), wolf (*Canis lupis*), red fox (*Vulpes vulpes*), beech marten (*Martes foina*), pine marten (*Martes martes*), badger (*Meles meles*), wild cat (*Felis sylvestris*), polecat (*Mustela putorius*), capercaillie (*Tetrao urogallus*), wood pigeon (*Columba palumbus*), partridge (*Perdix perdix*), and chukar partridge (*Alectoris chukar*). The eastern part of the range is well known for its birds of prey such as griffon vulture (*Gyps fulvus*), Egyptian vulture (*Neophron percnopterus*), and black vulture (*Aegypius monachus*) (Peev et al. 1995; Perry 1995). This wildlife also has attracted the attention of bird-watchers, local and foreign game hunters, and conservation organizations, as examples in later chapters illustrate.

As with other mountainous regions of the world, the species composition of vegetation in the Rhodope changes with altitude. The central and western portions of the range are heavily forested, and few areas are above tree line due to the moderate elevation. According to the director of the forestry directorate for the Smolyan region, which includes the village territory, 74 percent of the region's land is forested—making it the most forested region of the country—and more than 70 percent of these forests are coniferous.[5] Among them are found the southernmost

examples in Europe of boreal-type forests, which are composed
of subarctic coniferous species. The forests in the lower slopes of
the central Rhodope typically consist of deciduous species dominated by oaks (*Quercus frainetto, Quercus pubescens, Quercus pe-
traea*) at lower elevations and beeches (*Fagus sylvatica*) at the upper range. In some places, such forests were commercially logged
during the socialist period and replaced through reforestation activities with faster-growing coniferous species. Conifers also dominate the higher elevations. Dominant species are Norway spruce
(*Picea abies*) and Scots (or Scotch) pine (*Pinus sylvestris*), supplemented by Austrian pine (*Pinus nigra*) and silver fir (*Abies alba*)
(Bojinov et al. 1994; Georgiev 1993; Perry 1995:115–116). Above
and between the forests are farmlands, meadows, and pastures
with a wide mix of Central European, Mediterranean, and Balkan
flora (Perry 1995:115–116). The Rhodope Mountains rank second,
after the Balkan Mountains, in the number of plant species found
only in Bulgaria, with more than 80 endemic species and sub-
species. About 16 of these are found only in the Rhodope (Peev et
al. 1995:11).

In contrast to other mountain ranges in Bulgaria, no national
parks or other large protected areas have been designated in the
Rhodope, despite considerable landscape and biological diversity and interest in such a designation by national and international conservationists (e.g., Meine 1994:40–41; Peev et al. 1995).
The Rhodope Mountains were included in a short-lived proposal
in the early 1990s for a project called Ecological Bricks for Our
Common House of Europe, which sought to create a series of
transboundary national parks in Europe, straddling the borders
between countries of the former Eastern bloc and the West (Ecological Bricks 1990).[6] They are currently included in the Global
200 Ecoregions Campaign of the World Wide Fund for Nature
(Kemf 1997). The most recent developments in conservation planning are discussed in the conclusion of the book.

Meanwhile, numerous, smaller nature reserves and other protected sites have been designated in recognition of the region's
landscape and biological diversity. Among these are four small reserves designated under the Man and the Biosphere Program of
the United Nations Educational, Scientific, and Cultural Organization (UNESCO): Mantaritsa south of the town of Rakitovo,
Kupena near the town of Peshtera, Dupkata near the village of

Fotinovo, and Chervenata Stena near the village of Bachkovo. Other reserves and protected landmarks in the central Rhodope include Beglika, Buynovsko Gorge, Chudnite Mostove, Smolyanski Lakes, and Trigradsko Gorge (Georgiev 1993; Peev et al. 1995). *Chudnite Mostove* (the Wonderful Bridges) are three large natural stone bridges, located a short distance—two hours by foot or 13 kilometers by road—from Zaburdo. The bridges and the nearby modern chalet were frequently referred to by villagers in conversations about local landmarks, and tourists who have visited them sometimes pass through the village, occasionally stopping for groceries or a drink at a coffee bar.

The national and international status of these sites shows the region's conservation value, but such designations tend to be of lesser significance at the local level. As with other biosphere reserves in Eastern Europe, those in the Rhodope lack human populations and the zonation sometimes associated with biosphere reserves elsewhere (IUCN 1995:38; Martin Price, personal communication, 1998). A few sites, such as the Wonderful Bridges and the Trigrad Gorge, are identified by local officials and residents for their value as tourist attractions, and Chervenata Stena reserve is located along a busy regional road. Others, however, seem to have little meaning for the local residents. The Dupkata reserve, for example, is isolated from human settlements, and an official from the closest village told me that the reserve is of little consequence to his constituents.

Social and Cultural Setting: An Introduction to Zaburdo

The village of Zaburdo is located in the center of the Rhodope at the proverbial end of the road—albeit a paved one, wide enough for two cars to pass at all points (see Fig. 2). The nucleated settlement lies at an elevation of 1,200 to 1,300 meters above sea level, along a one-kilometer stretch of a small river. It is ringed on all sides by mountain peaks and surrounded at closer range by hay meadows, potato fields, steep rocky slopes, and coniferous forests.[7] Directly above the village, the north–south trending ridge reaches an elevation of 1,800 to 1,900 meters, and a few kilometers north along this ridge one encounters Golyam Persenk, one of the highest peaks in the range, whose tree-covered summit

Figure 2. Map of the central Rhodope

reaches an elevation of 2,091 meters. The village's upland location is reflected in the frequent use of the phrase, "I am leaving to go down" (*slizam na dolu*), when residents leave by road. The community's name is usually explained as a contraction of the phrase *zad burdo*, which means behind the ridge or hill and aptly describes the village's setting. Relative to other settlements, it is 13 kilometers from the main regional road, 28 kilometers by road from the municipal center of Chepelare, and an hour's drive from larger towns to the north or south.

Zaburdo's physical layout is constrained by its valley-bottom location and the steepness of the surrounding terrain. The paved main street parallels the watercourse, and a few secondary roads—some paved or lined with cement blocks—climb uphill from or run parallel to it. Houses are close together and have only the smallest of yards due to these space constraints. Many houses were built following agricultural collectivization, and most have electricity, telephone service, and indoor running water. Old-style houses in which living quarters are on the second story, above a ground floor sheltering the livestock, are rare. Instead, most households have separate wooden barns for their livestock, either adjacent to the house or at the upper end of the village in what used to be the agricultural yard of the cooperative farm. A few banks of garages are employed primarily for hay and potato storage, although they are sometimes used for sheltering cars, housing livestock, or buying wild-collected mushrooms.

Walking up the main street, one observes that a variety of basic services are available. Along with the mayor's office, the local administrative building houses the post office and branches of the postal bank and the state savings bank. Next door, the village health service is staffed by a doctor and midwife and has facilities for a visiting dentist. On the other side of the administration building is the *chitalishte*, an institution found in most Bulgarian settlements, whose name directly translates as reading room but is better characterized as a cultural center and library. Although the building's auditorium is infrequently used in comparison to the numerous concerts and film screenings during the socialist era, the library is open regularly, and I attended two holiday concerts plus a few community assemblies in the auditorium during my stay. Veterinary services are available at the upper end of the village, where there is also a privately owned milk-processing plant, which turns locally produced milk into cheese and butter. Children attend first a village nursery school and then a local school that goes up to the eighth grade. Now, many students subsequently attend high school outside the village, and more rarely they go to college.[8] Half-a-dozen village establishments serve coffee, soft drinks, and alcoholic beverages; however, the most that one could get to eat in these places during my stay was ice cream or packaged pretzels and croissants. When I arrived in December 1996, six or seven stores sold food and basic household goods.[9] But during the economic crisis in early 1997, all of them closed. Only two subsequently reopened, and later an additional new store opened. Nearly every village household frequents a bread bakery operated by the commercial cooperative, and notices about community events are placed on the bakery door with the knowledge that the information will quickly spread to residents.

Although not a large village, many settlements in the region are smaller. According to data from the village administration, 716 individuals lived in Zaburdo in early 1997 in 263 households, and national census data collected in March 2001 indicate a population of 669. (Most of the decline is accounted for by deaths greatly exceeding births.) Village population increases somewhat in the summer and on holidays when the town-dwelling children or grandchildren of residents arrive for vacations, to assist with agricultural activities, or both. In addition, a few town-dwelling households live in the village in the summer and engage in agricultural production on their own. Zaburdo's population is aging, with about one-third of the residents at or near retirement in 1996 (unpublished data, National Statistical Institute, 1998; Tsentralno Statistichesko Upravlenie 1988:113). This age structure may look extreme to those familiar with the youthful populations of the Third World; however, it resembles that of other Bulgarian villages on average, as shown in the comparative data in Table 1.

When discussed by outside observers in religious terms, Zaburdo is often identified as a Pomak or Bulgarian-Muslim village, although the terms are rarely used by residents. These terms are employed to describe people who speak Bulgarian as their first language but are Muslims or are otherwise associated with Islam through their names, kinship ties, history, or rituals (Todorova 1997:63). Bulgaria's Pomak population is concentrated in the Rhodope Mountains and the Mesta River valley, which borders the mountains on the west (Brunnbauer 1998; Konstantinov 1997; Silverman 1984). A common—but not the only—explanation for the origin of this group is that they are the descendants of ethnic Bulgarians who converted to Islam during Ottoman rule of this Balkan region (for additional information see Brunnbauer 1998; Eminov 1997; Konstantinov 1993, 1997; Silverman 1984; Todorova 1997; Zhelyazkova 1998). Many villagers related a widely cited story of forced conversion, even at knife point, which corresponds to tales of Turkish oppression. Several scholars question this story, and the situation in reality was likely more complex.

While some details of the lives of these people under Ottoman

Table 1. **Age structure of Bulgarian communities, 1992.**

	Total population		Urban population		Rural population		
Age	*Bulgaria*	*Smolyan*	*Bulgaria*	*Smolyan*	*Bulgaria*	*Smolyan*	*Zaburdo*
0–19 years	26.4%	30.2%	28.3%	33.1%	22.7%	27.2%	19.0%
20–59 years	53.1	55.4	55.9	56.7	47.2	53.9	53.4
60+ years	20.5	14.5	15.8	10.2	30.1	18.8	27.6

Source: Data from National Statistical Institute 1996b:2, 5, 8, 81–84, 218.

Notes: Smolyan data are for the Smolyan Region. Figures are the percentage of the population in each group. Urban population figures are for settlements classified administratively as towns (e.g., municipal centers), and rural population figures are for villages and similar small settlements.

rule are unclear, unambiguous evidence exists of efforts to manipulate their identities during the 20th century. In particular, they have been subject to as many as four assimilation campaigns designed to turn them (back) into pure Bulgarians. In Zaburdo's case, these efforts date back to the winter of 1912–1913, when the individuals carrying out the conversions wrote to church officials that they were prevented for some time from reaching the village due to deep snows and cold temperatures (Georgiev and Trifonov 1995:107). The campaigns sought to change people's names from Muslim or Turkic to Bulgarian or Slavic ones, to replace Islamic with Christian religious beliefs and practices, and to prohibit the wearing of clothing marked as Islamic (Konstantinov 1992). Besides assimilation campaigns, during the socialist era there were sporadic economic development initiatives in this mountainous border region, where the majority of Pomaks live, in an effort to improve local economic conditions through such things as lower taxation, credits for building houses, and the supply of foodstuffs (Konstantinov 1992:350; Todorova 1997). Konstantinov (1992) describes such improvements as a "carrot" against the "stick" of name-changing campaigns; however, it is also the case that conditions in these rural mountain communities were generally less developed than in other parts of the country.

Although residents of some Muslim communities in Bulgaria have rebuilt mosques and returned to Muslim names in the post-socialist period, such is not the case in Zaburdo.[10] Religion played little role in everyday life during my fieldwork, and there is neither a church nor a mosque in the village. There is also no support for the so-called Turkish or Muslim political party, the Movement for Rights and Freedom. When asked about religion, village residents surveyed most commonly answered that they were atheists. About one-third of surveyed households identified with Islam, but some of them added that this was because of their ancestry and that they are not practicing Muslims. Others reported feeling more inclined to Christianity, and some of them, but not all, were Christians who had married into the village or were descendants thereof.[11] In a nutshell, religion plays little role in the identity of most villagers; for most villagers, what is important is being Bulgarian.

This aspect of village identity remains significant, however, because religion played a greater role in the presocialist and, to a lesser extent, socialist eras. Speaking to older residents, one learns that the first village schools were religious ones, that there used to be a mosque, that mixed marriages were frowned upon, and that women formerly wore veils. Some of the first women to stop wearing veils did so at the insistence of husbands who were communist party supporters. Of particular interest for the issues discussed here, this religious identity likely affected demographic patterns and economic strategies in the past, the results of which are still seen today. Reportedly the residents of such villages did not participate in the large-scale migration from village to town that characterized much of Bulgaria and Bulgarians in the post–World War II period, insofar as they could avoid it (Bates 1995; Konstantinov 1992:356, 1993:69). Instead, they remained in their villages, or in some cases moved to other rural locations, where they worked on cooperative farms and in small-scale industrial units or rural assembly workshops (Brunnbauer 1998:2; Konstantinov 1997:42–43). In the postsocialist period, the tendency to remain in rural settings has not necessarily been an advantage for these communities, because the regions inhabited by Pomaks have been severely affected by the economic crisis and accompanying liquidation of the cooperative farms and closure of the small, village-based industrial enterprises (e.g., Konstantinov 1997:43; Todorova 1997).

Meanwhile, in political terms, Zaburdo is sometimes referred to as a red, or socialist, village. The incumbent communist-turned-socialist mayor was reelected after the political changes and served until he died in an automobile accident shortly before my arrival in 1996. Zaburdo was the only settlement in the municipality where the socialist party received the most votes in the 1997 parliamentary election. While support for the socialist party is not uncommon in Pomak villages (see Brunnbauer 1998) and for that matter in rural settings more generally (Creed 1995c), subsequent elections have called into question the strength of this identification. The former mayor's son-in-law, a local entrepreneur, was elected as an independent to serve out his father-in-law's term in a special election in 1996 and then reelected in the next regular mayoral election in 1999, this time with backing from the anticommunist UDF. Also, like the rest of the country, the former king's party received considerable support in the village during the 2001 parliamentary election. Some people maintain socialist-party loyalty and continue to vote "red," but others say that party affiliation is less important than the candidate and what he or she can do for the village and its residents. While the formal authority of the mayor is more limited than it was during the socialist era, and funds are tight, the office continues to play an important, traditional role in dispute mediation that avoids the expense and time associated with court action. Villagers often approach the mayor's office with problems, whether regarding disputes between individuals or assistance with aid programs.

Changes in Zaburdo's population over the last century (see Table 2) reflect larger political and economic influences in the Balkans alongside the natural processes of birth and death. The significant drop between 1910 and 1920 corresponds to a period of substantial disruption in the region, including relocation of the international border between Bulgaria and Turkey to the south in 1912 and a subsequent religious conversion campaign in 1912–1913 (Georgiev and Trifonov 1995). Many people left in search of religious freedom and economic opportunities.[12] Population growth subsequently resumed, peaking in the mid-1950s, just before creation of the cooperative farm in Zaburdo. This event produced another wave of outmigration, as people left in response to loss of their land and livestock and the limited employment opportunities on the farm. In addition, birth rates dropped after 1963, while death rates remained the same (Hadjiev 1969:16). This birth-rate decline may reflect outmigration by young people—that is, potential parents—following establishment of the cooperative farm, although it also mirrors a longer-term, nationwide pattern of decreasing births (McIntyre 1980). The postsocialist

Table 2. Zaburdo population, 1880–2001.

	1880	1887	1892	1900	1905	1910	1920	1926
Population (1)	850	636	550	1,027	1,083	1,202	641	834

	1934	1946	1956	1965	1975	1985	1992	2001 (3)
Population (2)	967	1,090	1,151	1,070	931	865	779	669

Sources: (1) Unpublished data card for the village of Zaburdo from the head directorate of the census. Located at the National Statistical Institute, Sofia; (2) National Statistical Institute 1994:143; and (3) unpublished data obtained from the Zaburdo village administration.

Notes: Data up through 1956 are for people present. Data from 1965 onward are for people registered at the location.

period has similarly seen further population decrease through a combination of lowered birth rates, death rates exceeding birth rates, and continued outmigration. Between 1990 and 1997, for example, 95 people left the village, while only 15 people moved in (unpublished data, National Statistical Institute and Zaburdo administration). This refutes the popular "return-migration" hypothesis, which states that, as employment opportunities in the cities decreased, people would move in large numbers to villages where they could at least feed themselves by working the land (see also Creed 1998:261). A few couples at or near retirement age have returned to live in Zaburdo and farm the land restored to them, and other individuals periodically return to the village to cultivate the restituted land without moving to Zaburdo. The number of young people returning to live in the village is very limited, however.

Ecology and Economy in Historical Perspective

In broad outlines, Zaburdo's history can be divided into three categories. The first is often referred to by villagers as "private times," by which they mean the period before the collectivization of agriculture. In political terms, the second period, "during communism," refers generally to the period between September 9, 1944 and November 10, 1989, although creation of a cooperative farm in the village in the late 1950s and its subsequent dismantling in the early 1990s were perhaps more significant events for the daily lives of villagers. The remainder of this chapter focuses on these two periods, while the following two chapters address the postsocialist period. Although some sources cite evidence of the village dating back more than 600 years, these data are limited to occasional citations of documents with references to a village assumed to be Zaburdo because of the similarities in name and general location. Consequently, my discussion focuses largely on the period after Bulgaria gained independence and for which the data are more clear. It shows how Zaburdo and its residents were increasingly incorporated into a modern cash economy over the last century and the ways in which larger political and economic forces affected resource use and livelihood strategies before 1989. It also provides a baseline against which to view the changes in natural resource use and management under the postsocialist conditions of agricultural land restitution and rural economic reorganization—changes that are the topics of subsequent chapters.

During "Private Times"

Bulgaria's autonomy from Ottoman rule is generally dated to the signing of the Treaty of San Stefano on March 3, 1878, although full independence did not occur until 1908, and the southwest corner of the country remained under Ottoman control until 1913 (Lampe 1986). Bulgaria emerged from five centuries of foreign control as a predominantly agricultural country of rural peasant smallholders with little industrial base (Crampton 1987; Lampe 1986). Accounts of agriculture at the time often talk of the small size of landholdings, their dispersal in numerous separate plots, and the low level of technology employed (e.g., Crampton 1987; Sanders 1949; Stillman 1958). The limited industry of the early 20th century was dominated by food processing and textiles, largely dedicated to processing the raw materials from Bulgaria's rural agricultural base (Lampe 1986:69).

Following Bulgarian independence in 1878, Zaburdo was included in Bulgarian territory. The international border at the time was only 6 kilometers to the southwest, however (Karov 1993; Iliev 1973:53), and some neighboring villages where residents had trading and kinship relations remained on the Turkish side of the border for another three decades. Village services and public infrastructure were limited. Reflecting Zaburdo's Islamic past, a "Turkish" or Islamic religious school was established as early as 1867 (Siuleymanov 1968:57). When Dimitur Karov (1993:8) arrived from neighboring Shiroka Luka to serve as the village secretary in 1901, there were as many as five small prayer houses that were also used as religious schools for children. The last of these Islamic schools closed in the 1940s due to a lack of teachers. The first secular Bulgarian teacher arrived in 1903, and a public school was established in 1907–1908, bringing the possibility of literacy in Bulgarian to residents. Zaburdo was reportedly one of the first Pomak villages in Bulgaria to have a secular Bulgarian school, and the fact that the village is less religious or more progressive than other Pomak settlements is sometimes attributed to this development. Initially, the school only provided classes up to the third grade, but the number of grades the school offered increased over time. Classes were open to girls as well as boys, provided their parents sent them. This was not always the case, as my conversations with older grandmothers revealed, and boys were sometimes discouraged from continuing their educations outside Zaburdo because their parents wanted them to stay in the village and help their families with agricultural production.

Even though the economy during this period was largely subsistence oriented, village residents had increasing contact with Bulgaria's political and economic system between independence and World War II. By the early 20th century, a few coffee shops or small private stores sold basic supplies such as sugar, salt, and kerosene. Travel to and from the village, including the transport of goods, took place by foot or with pack animals over a network of trails, as there was no vehicle road to Zaburdo at the time. A road along the north–south axis between Plovdiv, Asenovgrad, and Chepelare was not built until 1911 (Karaivanov 1990:42), and even this was still some distance from Zaburdo. Monov (1987:72) cites lack of transportation infrastructure as an important factor in the limited economic development of the Rhodope region generally up through the first half of the 20th century. Economic opportunities in the village were largely restricted to farming and animal husbandry.

Precollectivization Landholdings and Land Ownership

The broad outlines of precollectivization property relations in Bulgaria began to emerge in the mid-19th century. In the context of changes in the degree of Ottoman control over the region and the Ottoman land reform law of 1858, individual farmers and households gained more formal control over the land that they worked (Crampton 1987; Islamoglu 2000; Jorgens 2000; Stillman 1958; cf. McGrew 1985). Nearly five centuries of Ottoman rule had essentially obliterated claims to large estates of land. Thus, following independence in 1878, Bulgaria was primarily a country of independent smallholders who farmed land that they owned (Lampe and Jackson 1982:135). These agricultural landholdings tended to be small, and a given landowner usually owned several plots scattered in different locations. Stillman (1958), for example, lists an average farm size for Bulgaria of 7.3 hectares in 1897 and 4.5 hectares in 1946. Similarly, Crampton (1987:137) writes that average Bulgarian landholdings in 1926 were 5.7 hectares, divided into 16 strips, and that by 1946 the average holding had fallen to 4.3 hectares divided into plots averaging 0.4 hectares. One reason often cited for the fragmentation and decline in holding size over time is the tradition of partible inheritance, in which land is divided equally among the sons and sometimes also the daughters of a landowner (Crampton 1987:137, 167; Lampe 1986; Stillman 1958; cf. Bishop et al. 1994:77; Cole and Wolf 1974). Land reform efforts were carried out in 1920–1921 and again after World War II with the intent of limiting holding size and redistributing excess land to those with little or none, but they had limited impact in large part due to the relative absence of large holdings (Crampton 1997; Lampe 1986:57; Stillman 1958:76; Strong et al. 1996:19). This reinforces the point that most agricultural landholdings in Bulgaria before collectivization were small with few large estates.

Literature on the central Rhodope similarly suggests that local residents, both Christians and Muslims, gained ownership of private land during the latter half of the 19th century. In some cases, agricultural lands were created by burning forests, although forestry authorities prohibited this practice in the late 19th century (Karaivanov 1990; Primovski 1973). The general Bulgarian pattern of small and fragmented landholdings applies to the Rhodope region, as well, although holdings there were often smaller than the national average. Monov (1987:62), for example, indicates that average agricultural landholdings in the Smolyan and Asenovgrad regions in 1937 were 1 to 2 hectares, and the average amount of agricultural land restored to former owners in Zaburdo (i.e., their precollectivization holdings before division among heirs) was 2.7 hectares (unpublished data, Chepelare Municipal Land Commission, 1997). As elsewhere in Bulgaria, women could own and inherit land, although in practice they often did not inherit as much as men.[13] Some forest land was also in private ownership in the region, and some pastures and forests around the village were managed as community property. In some cases, the local communities had use rights to but not outright ownership of these lands, which instead remained in the hands of the state. This has become an issue in the postsocialist restitution of forest lands, when local communities claim lands that they consider theirs but for which they lack documents of outright ownership. As will be elaborated upon later, private forests were quite common in this region, and many of them were managed cooperatively in the 1920s to 1940s. Thus, precollectivization property relations in Zaburdo were characterized by private ownership of more intensively exploited agricultural lands and communal or joint ownership or management of more extensively used forests and pastures (Netting 1981 describes a similar pattern in a valley in the Swiss Alps).

Agricultural Activities during Private Times

Before collectivization, agricultural activities in Zaburdo were primarily subsistence-oriented and organized on a family or household level. For tasks requiring more people than were available within the household, such as harvesting or threshing grain, traditions of mutual labor exchange existed. Production took place using simple tools like hoes and wooden plows, animal traction and conveyance, and natural manure fertilizer. Manuring was sometimes accomplished by pasturing sheep in the fields, with the amount of time spent in a particular field depending on the number of animals a landowner had in the flock. Several villagers told me that this practice resulted in productive fields. Manure was also transported from barns to the fields using pack animals. Crop production and livestock raising were closely tied. Manure was necessary for maintaining soil fertility and thereby making possible continuous use of the fields, and the livestock were fed some of the crops or their byproducts (Vakarelski 1969:51). Yet, several observers note that production in the central Rhodope was low due to the poor quality and limited quantity of land (e.g., Monov 1983; Primovski 1973; Siuleymanov 1968:49, 64).

Precollectivization crop diversity in Zaburdo went beyond the potatoes, beans, and garden vegetables grown today to include lentils, wheat, rye, oats, barley, spelt (*Triticum spelta*, a hardy grain that is distantly related to wheat), flax, and hemp (Smolyan regional archives, Fund 894, op. 1, a.e. 14). Indeed, when asked about plants and animals that had disappeared, most people responded with the names of these domesticated plants along with oxen, instead of the wild flora and fauna I had expected. Grains were used for both animal fodder and human consumption, and hay was produced to feed the livestock. The area of hay meadows then was smaller than in the postsocialist period, however, because more fields were planted with these other crops. Flax and hemp were used in cloth production, an activity that occupied women during winter months (Dechov 1903; Gumnerova 1969). Primovski (1973:264) suggests that the cultivation of garden vegetables is a relatively recent innovation in the central Rhodope, beginning only in the last hundred years, and that the variety of vegetables seen today did not emerge until the 1920s. During private times, potatoes were grown as one crop among many and in smaller quantities than today. Potatoes originated in the Americas and are now grown in many mountain regions around the world; exactly when they arrived in the Rhodope is unclear. The first potatoes may have been grown in Rhodope mining communities around 1830 (Primovski 1973:268), and Iliev (1973) suggests that potatoes were being cultivated in Zaburdo by the end of the Russian-Turkish war of 1878.

Although agricultural production was predominantly for subsistence, some exchange of products occasionally took place between the village and the lowlands and with other Rhodope villages. Some local residents reportedly produced enough wheat to sell to neighboring villages. A villager born in the early 1940s related another example to me. Her father, who was described as a successful farmer with substantial land and livestock holdings, transported some of his produce by pack animal to lowland villages. There he exchanged his produce for staples and agricultural products that could not be grown in Zaburdo, such as grapes, tomatoes, and cabbage, or he sold his produce for cash and then used the money to purchase the other goods. He did not sell potatoes, she said.

Precollectivization Animal Husbandry

Although residents engaged in agriculture, it is animal husbandry for which the Rhodope region is best known. The historical importance to the region's economy of livestock raising, particularly sheep production, can be seen in the attention given to this activity by Bulgarian historians and ethnographers such as Dechov (1968), Kanev (1975), and Primovski (1973). In Dechov's (1968:198) classic description of sheep breeding in the central Rhodope, originally published in 1903, Zaburdo is included in a list of villages in which most residents raise sheep. The livestock varieties raised were typically local ones adapted to the mountain environment. The large number of small stock—especially sheep, but also goats—suggests that some animals were raised for the market, while the production of other livestock varieties was often more oriented toward household provisioning. Until the 1920s, it was common for sheep to be owned in large numbers in the Rhodope, and only the poorest families raised a small number of sheep (Vakarelski 1969:55). Primovski (1973:313–314), for example, lists a handful of individuals in Zaburdo with 300 to 1,000 sheep and writes that many villagers had 10 to 15 sheep around the time of independence. In terms of overall numbers, there may have been as many as 18,000 sheep and 6,000 goats in Zaburdo before Bulgarian independence, there were about 16,000 sheep and 3,000 goats until many residents left the village in 1914, and by 1928 the numbers had declined to 8,000 sheep and 2,000 goats (Primovski 1973:314). Thus, early in the century, Zaburdo residents owned considerably more animals than they owned in the mid-20th century, when they only held 3,300 sheep, 400 cattle (including about 100 oxen), and 100 or so goats (Smolyan regional archives, Fund 894, op. 1, a.e. 14 and 16). (Currently, small-stock numbers are even smaller, and oxen have disappeared.)

These animals were often herded by someone regularly paid with money or produce to do this job, and a few older men in the village describe their occupations as shepherd or other type of herder. They finished the third or fourth grade in the local school and spent the rest of their working lives with animals, first as private herders and then later as employees of the cooperative farm. Herding animals was typically an activity for men and boys, and ethnographic museums and books often have displays or pictures of the clothing and equipment involved. Girls or women occasionally went out with the animals around the village, as I learned one day when an 80-year-old grandmother told me of her encounter with a wolf while out herding as a young woman.

Sheep production was often a seasonally transhumant activity. The sheep and the shepherds charged with their care spent summers high in the mountains, sometimes in seasonal camps outside Zaburdo and other villages. The sheep were milked during this season, and milk-processing facilities (*mandri* or dairies) were located near the flocks. The milk was made into products that were more easily stored, such as a white feta-like cheese, a yellow cheese called kashkaval, butter, and yogurt. Flocks often spent part or all of the winter season in the lowlands, which meant less need to secure animal food for the cold, snowy mountain winter. Some people with smaller numbers of sheep wintered them in the mountains, while others drove their flocks to lowland areas for the second half of the winter. When an individual owner had 1,000 or more sheep, it was difficult to keep them near the village for more than a month at a time. These larger flocks were taken south to the Aegean coast for the entire winter and spent all summer at dairies high in the mountains (Dechov 1968). Use of Aegean pastures was possible in part due to widespread mosquito infestation of the coast during the summer and the resulting risk of malaria, which limited year-round settlement there (Forbes 1997:195). As a result of losses in World War I, however, Bulgaria signed the Treaty of Neuilly in November 1919 in which western Thrace was transferred to Greece, and access to the Aegean was essentially precluded after this date (Jelavich 1983). Rhodope villagers consequently lost a market for their production, and the loss of access to fertile winter-grazing lands

prompted sheep owners to reduce the sizes of their flocks (Monov
1983:7; Primovski 1973; Siuleymanov 1968:49).[14] Reduction in
livestock numbers also had implications for field fertility, because
less manure was available for use as fertilizer (Iliev 1973:75).

Raising work animals and cows was, in comparison, a smaller
enterprise. Many village households owned a cow or two, primarily for milk and secondarily for meat for household consumption. The cows were pastured around the village during the summer. Oxen, horses, donkeys, and mules were used as draft
animals for agricultural production, and the latter three also were
used as pack animals. Mules were popular pack and draft animals because of their stability on steep, rocky slopes and paths,
and the poorest families owned donkeys (Primovski 1973:346).

Forest Management and Exploitation of Other Resources

Rhodope residents have long relied on forests for fuelwood for
heating and cooking, for building materials, and for other subsistence uses. Most houses are built of stone or brick, but wood is
also used in their construction, for beams, for example. Local men
hunt, and villagers also collect wild fruits and herbs for tea and
medicinal purposes. A 1939 article in the Chepelare newspaper
Rhodopska Iskra discusses the potential contribution of wild herb
and fruit collection to the income of mountain residents, reporting that there was an international market for several plants that
grow in the region (Organizirane iznosa 1939:2–3).

Yet, forests did not become an object of significant commercial
exploitation until after Bulgarian independence, in part because
of the limited transportation infrastructure that existed until the
20th century (Primovski 1973:352). As access to markets and pastures in the south was increasingly difficult following independence and then impossible after the 1919 border closure, residents
of the central Rhodope turned toward the forests as an alternative
income source. Initially, wood for heating was transported by
river to the Plovdiv area. Later, as roads and other infrastructure
for the timber industry were put in place, the forests were commercially exploited for lumber (Karaivanov 1990:23).

In the central Rhodope, as elsewhere in Bulgaria, there were
private, municipal, church, mosque, school, and state forests in
the past. Overall in Bulgaria, about 16 percent of forests were un-

der private ownership before their nationalization in 1947 (Stoyanov 1968). In the Smolyan *okoliya* (a former regional administrative division), however, about 54 percent of the forests were private in 1945 (Monov 1987:75), and for some central Rhodope communities the share of private forests may have exceeded 80 or 90 percent. This high percentage is due to the importance of forest exploitation for the livelihoods of the region's residents, along with the availability of opportunities to purchase forests.

Initially, after Bulgarian independence, people with use rights to forests essentially retained ownership of them and in some cases the forests had been purchased. (At the time, use rights to forests as sources of pastures and watering locations for livestock were perhaps more important than forests as sources of timber.) In 1904, however, the Bulgarian parliament decided that there had not been privately owned forests before the Ottoman period—only use rights—and the government took control of the forests. This occurred despite protests from local residents (e.g., Iliev 1973:58). Divested of their forests, some people left the region. To combat the poor economic conditions in and depopulation of this strategic border region, the parliament later reversed its decision, passing a law in 1911 that provided for the sale of forests to residents of specified settlements in the Asenovgrad region, including Zaburdo. Although implementation of the law was delayed by the Balkan Wars and World War I, many people took advantage of this opportunity, buying forests and exploiting them commercially. By 1927, for example, there were 54 sawmills in the Chepelare area, including two in Zaburdo (Primovski 1973:358). Exploitation of these private forests was supervised by the state, which regulated, for example, how much could be cut in a given location.

In addition, legislation passed in 1923, during a period of agrarian rule in Bulgaria, required that private forest holdings in the Rhodope below a certain size be managed cooperatively (Karaivanov 1990; Monov 1987:75).[15] Such private forestry cooperatives were best established in the central Rhodope municipality of Chepelare. A forestry co-op was created in Zaburdo in 1931 to assist with forest management and the exploitation and sale of timber. Besides providing forest owners with income, these cooperatives donated money to local development projects. Zaburdo's cooperative provided money to build the village school and municipal building in 1939, for example, and later on, it provided financial assistance for road construction (Kuzmanov 1967:101; Smolyan regional archives, Funds 372K and 373K). Such private forestry cooperatives from the presocialist era continue to enjoy popular support even today because people associate them with providing infrastructure to local communities, income for forest owners, and employment opportunities.

Economic Life under Socialism

Soviet troops entered Bulgaria's capital city of Sofia on September 9, 1944—the date generally celebrated before 1989 as the beginning of the socialist era. This event was followed by the consolidation of power in the Bulgarian Communist Party, culminating in the declaration of a "People's Republic" in December 1947 (Crampton 1987:145; McIntyre 1988:16). The socialist development plan in the postwar period brought industrialization, urbanization, and the collectivization and modernization of agriculture, significantly changing the structure of the economy. A Soviet-style, centrally planned economy was adopted, with state ownership of the means of production and party guidance of economic affairs, and private industries were nationalized (McIntyre 1988:79, chap. 4). The proportion of the population living in urban areas rose as the economy industrialized, from about one-quarter at the end of World War II to about two-thirds in the 1990s (National Statistical Institute 1996a:44–45, 1998). The period was also marked by heavy involvement in trade, primarily with other socialist countries through the Council for Mutual Economic Assistance (McIntyre 1988).

Agricultural activities in Bulgaria were also substantially affected by socialist-era economic reforms, particularly in terms of the organization of production and the level of technology. In contrast to the nearly universal private ownership of almost one million small family farms at the end of World War II, by 1960 virtually all agricultural land was controlled by either village-based cooperative farms or a few state farms. Further consolidation of agriculture took place in the 1970s with the merger of cooperative farms into a smaller number of larger agro-industrial complexes (Dobreva 1994:340–341; Meurs and Djankov 1998; Stillman 1958). Instead of peasant smallholders, wage laborers provided the labor power for agriculture, and people not needed in cooperative production often moved to the towns in search of year-round employment in industrial enterprises. As people moved away, the declining rural population was compensated for in agriculture by increased mechanization at the large state-run farms and increased use of modern agricultural inputs such as commercial fertilizers and pesticides. Even so, the level of government investment in agriculture was low compared to other countries in the region, with comparatively more resources going into the nation's industrial development (Meurs and Djankov 1998:48–49).

This development plan resulted in overall economic growth, particularly in the early years (e.g., Lampe 1986; McIntyre 1988; also Meurs and Djankov 1998 on agriculture), and there was also some material improvement in people's lives. Sanders (1949:32), for example, laments infant mortality rates of 123 per 1,000 in 1941, comparing them to 46 per 1,000 in the United States in the same year. By the late 1980s, the infant mortality rate in Bulgaria had dropped nearly tenfold to 14 per 1,000 (National Statistical Institute 1996a:70). Similarly, education opportunities improved for people in both rural and urban areas, such that today education levels compare favorably to countries in the west. The illiteracy rate, which had declined to 31.5 percent of the population over 7 years of age in 1934, was further reduced to 8.3 percent in 1965, one of the lowest in the Balkans. In 1992, 10 percent of the adult population held college degrees, compared to 5 percent in 1975 and 1 percent in 1934 (National Statistical Institute 1996b:32). While much of the attention was directed to urban settings, rural communities also saw benefits in terms of transportation and communication networks, nonagricultural employment opportunities, health care services, the introduction of a social safety net, and increased educational opportunities. (I would be remiss in not pointing out that the system that generated these benefits was ultimately unsustainable.)

Socialist Village Life

Beginning in the late 1940s, Zaburdo residents saw numerous changes in their lives, including the material improvements just described. A vehicle road constructed in 1949 connected the village with the main regional road, and regular bus service started around 1953–1954, linking Zaburdo first with Plovdiv and Asenovgrad and later with Chepelare and Smolyan as well. The 1950s also saw the opening of a village health service, the installation of a village water supply system, electrification, and the establishment of a bakery by the commercial cooperative, eliminating for villagers the regular chore of baking bread. The current cultural center was built in 1966, and the health service building was constructed in 1971 (Iliev 1981).

These improvements and the "easier" life they had as workers on the cooperative farm or in other enterprises were frequently cited by villagers, who said that life was better under communism—better than private times and better than the postsocialist situation. They worked fixed shifts on their jobs and did not have the worries of private production, such as obtaining agricultural inputs and selling their crops. There were things to buy in the well-stocked stores, and they had money from their wages to buy what they needed. And not only food; they said they were also able to build and furnish new houses and put their children through school. Bread from the bakery was so inexpensive, a few cents (*stoltinki*) a loaf, that they would use it straight from the bakery—and thus still warm—to feed their livestock, then throw it in the river when it got stale. Sometimes they would tell me that their village was referred to as "little America" because they had everything they needed and life was so good.

A reality check is appropriate here. In this village I rarely heard stories about the less positive aspects of socialist rule in Bulgaria—children who could not find jobs or attend university because they came from rich or otherwise discredited families, fears that telling political jokes in the wrong company could result in being sent to political prison camps, restrictions on international travel, violence associated with religious assimilation campaigns or even relocation of some Pomaks from villages on the Greek frontier, shortages, and other bizarre aspects of an economy where prices were not associated with supply and demand. (All of these examples come from Bulgarians I met and spoke with elsewhere.) In part this may stem from the contextual or relativist nature of how people consider the past: Their nostalgia for their "better life" under socialism, particularly economically when compared to the present difficulties, puts these more negative aspects of the socialist system in background. There are, of course, some Zaburdo residents who comment on these issues or explain their preference for the present in terms of greater freedom of choice or the unfairness of collectivization, but they seem to be in the minority in the village.

Agricultural Production under Collectivization

Collectivization of agriculture in Zaburdo began in 1958–1959, with the creation of a labor cooperative agricultural enterprise (*trudovo-kooperativno zemedelsko stopanstvo* [TKZC], and referred to in this work as the "cooperative farm"). The cooperative farm was created at the end of collectivization efforts in Bulgaria, with lands in the more productive plains areas being collectivized first. Besides agricultural land, livestock and agricultural tools were taken from residents in the collectivization process. Villagers changed from independent, largely subsistence farmers to cooperative farm employees who exchanged their labor for money, which they could then use to buy goods from village stores.

Some people were resistant to the idea of collectivization, at least at first, and in some cases village farmers were "persuaded" through social pressure (and occasionally more) to join the cooperative farm. One man reportedly left with his family for several years, rather than join the farm, and another was said to have been forced to sign papers to join. As well, during the farm's early years, many people left the village for towns in response to losing their productive resources and the uncertainty about job availability—particularly year-round employment. Most agricultural work on the cooperative farm was seasonal, with only limited winter jobs in livestock production. Thus, forced collectivization was not necessarily a happy event for all, especially the larger, more successful private farmers. A former farm official estimated that 30 to 40 percent of the village's young people left at the time, starting a population downturn that continues. Some individuals who left the village later returned—some a few years later, others following the postsocialist restoration of private agricultural land. But other people were reportedly enthusiastic about the farm. They had no money with which to enlarge their private holdings and thus increase their incomes through private agriculture,

according to a former farm official. By working together on the cooperative farm, they could increase the amount earned, and their
earnings took the form of money rather than products. Eventually, nearly all households remaining in the village joined the
farm, and even the few who did not join worked on it.

The cooperative farm was involved in both crop production
and animal husbandry. At first crops such as tobacco and rye were
planted, but soon the farm began to specialize in potato production and, after 1965, in the production of seed potatoes. The
switch to monocrop production of potatoes likely reflects a combination of ecological and economic factors. With the creation of
the farm, the emphasis switched from production primarily for
subsistence to production for commercial distribution or sale. As
Cole and Wolf (1974:84) suggest, a shift in production, from crops
to satisfy household needs to crops for market sale, may be accompanied by rationalization of production. Farmers—or in this
case cooperative farm managers—grow that which can be produced most profitably and thus those crops that represent the
maximum productive use of the land. While use of such market-
oriented terms may seem odd in the context of collectivized agriculture, potatoes are a productive crop in mountain settings in
terms of calories per unit of land area (Netting 1981:163; Stevens
1993:79). The farm also produced strawberries, raspberries, and
black currants briefly in the late 1960s. Villagers speak fondly of
these days and of their first view of a helicopter, which airlifted
the fruit from the fields to the city. Finally, sheep and cows were
raised for both milk and meat. The farm produced hay to feed
them, and concentrated animal feed was obtained from grain-
producing enterprises in the plains because such crops were no
longer grown in Zaburdo. The precollectivization pattern in
which flocks of sheep spent January to late May in the plains near
Plovdiv continued, thereby reducing the quantity of hay and forage otherwise needed for the winter.

With the collectivization of agriculture, labor organization and
production technology changed. Agricultural labor was accomplished through brigades rather than households. These brigades
and their subdivisions were generally organized on a neighborhood basis, with one brigade each for the upper, middle, and
lower parts of the village. In addition, the specialization of agricultural labor increased. For example, while women continued as

general agricultural workers, many men became specialized as "drivers" or "tractor drivers" (see also Creed 1995a on this gendered division of labor).

Production technology initially did not change significantly from that employed earlier by private farmers. Lack of roads hampered mechanization, and land was consequently worked using the simple wooden plows and draft animals (horses and mules) that had been expropriated from village residents. Oxen were deemed unproductive; they were only useful for plowing in the spring but had to be fed year-round, and consequently they disappeared from agricultural production. Road building for agricultural purposes started in the early 1960s, and by 1962–1963 trucks could be used to transport goods and people, and tractors could be used for working the land. A former tractor driver for the farm related that every few years he received a new tractor, and his old tractor was given to less senior drivers, which suggested some level of investment in agriculture. Although the trucks were also employed to transport manure from cow and sheep farms to the fields, use of artificial fertilizers as well as modern pesticides, herbicides, and fungicides increased. Pesticides were applied to combat the Colorado potato beetle (*Leptinotarsa decemlineata*), which villagers report arrived in the late 1970s or early 1980s, and fungicides were used against potato late blight (*Phytophthora infestans*).[16] These problems and the consequent use of chemicals may have emerged with sustained monocrop production of potatoes on the cooperative farm.

Organization of the farm evolved over time, as was the case for all agricultural production in Bulgaria. Specifically, the village-based farm was first incorporated into increasingly large production units. Later it regained greater independence, as the productivity of the larger units decreased in the 1980s. These administrative changes were primarily made above the village level, although they affected who made production decisions. Some villagers formerly in management positions on the farm spoke of these consolidations critically. They commented that removing decisions from the purview of the villagers, who knew the land, and placing the responsibility at higher levels had a negative impact on production. They also said that the added administrative levels were a further drain on resources.

Another change in the organization of production took place in

the late 1980s (see also Creed 1995a). This was done because of a
lack of interest in agricultural work and declining agricultural
productivity. The novelty of the cooperative farm had worn off,
and people were moving out of the village, transferring to nonagricultural jobs such as those available in village-based assembly
workshops (discussed below), or retiring. The number of potato
production brigades had declined from three to one, and new
means were sought to draw labor into agriculture.[17] The new production method, called the family *akord* system, resembled contract farming. Its basic goals were to increase the number of
people involved in agriculture and the amount produced. In
Zaburdo, this system was employed primarily in potato production. Groups of village residents were allotted land to cultivate.
The cooperative farm provided the mechanized cultivation and
transportation services necessary for working the land, as well as
inputs such as seeds and fertilizer, while the groups of village residents were responsible for the hand labor. The resulting potatoes
were marketed through state channels, so producers did not need
to find buyers, and the income was divided among producers, the
farm, and the state.

This system was reportedly effective. Many villagers decided
to participate, even teachers, forestry enterprise workers, and assembly workshop employees who were not employed by the farm
on a regular basis. As much as 80 to 90 percent of the farm's land
was worked in this fashion. Production increased, as the producers were more conscious of costs and their earnings depended on
how hard they worked, providing an incentive to work harder.
Several villagers interviewed in the postsocialist period identified this system as the best of both worlds. They were paid based
on how hard they worked and thus could earn more money by
working more, but at the same time they did not have to worry
about things like getting the fields plowed, obtaining and transporting seeds and fertilizer, and finding a market for their produce. As will be seen, these are particularly problematic aspects
of postsocialist production, which likely contributes to the nostalgia for the *akord* system.

After collectivization, Zaburdo households, as with other
households in Bulgaria, were also allowed to engage in a limited
amount of personal production. Each household could keep a
donkey, a cow, a calf, and up to five sheep and work up to 5 de-

cares of land allocated to them from the cooperative farm's land.[18] Later the amount of land was increased to 6 decares for farm employees, and the number of permitted animals increased somewhat after 1975–1976. After 1970, the farm in Zaburdo would buy animals from private individuals, and in exchange producers had access to reduced-price concentrated forage. Villagers could rent mules or tractors from the farm for use in working these plots, as well as purchase agricultural inputs such as fertilizer. Thus, the farm facilitated several aspects of this personal production.

In some parts of the country, such personal plots were the source of substantial produce and livestock that eventually made their way into the formal distribution system. Crampton (1987:197), for example, reports that in 1982 personal plots covered only about 13 percent of the arable land in Bulgaria but produced about 25 percent of agricultural output, including 26 percent of fruit, 33 percent of vegetables, and 51 percent of potatoes (see also Lampe 1986).[19] Personal plot production for sale was not a priority in Zaburdo, however. One woman, for example, explained that her family did not sell the produce from these plots because land was limited. They might have one decare of potatoes, she said, and the remaining land was used to produce hay to feed the animals. My impressions are similar to those of Creed (1998:191), who observed relatively infrequent sales of personal-plot production in Zamfirovo in the late 1980s. He also cites figures for Bulgaria as a whole indicating that such production was not as widespread in terms of the numbers of households as might be suggested by the overall production statistics cited above. In the case of Zamfirovo, villagers explained their lack of sales in terms of the poor quality of the land (e.g., rocky soil, unreliable moisture). Creed suggests other factors in the lack of a commercial orientation in private production—involvement in nonagricultural enterprises and the *akord* system, along with the importance of private-plot produce in informal exchange relationships. Similar circumstances likely existed in Zaburdo.

Forest and Wild Resource Exploitation during the Socialist Period

Throughout Bulgaria, forests were nationalized in 1947–1948 (Bojinov et al. 1994; Karaivanov 1990). With this change of ownership, forests were managed and exploited through locally based, state-controlled forest enterprises, which were sometimes affiliated with cooperative farms, and many villagers were employed in these enterprises in both logging and reforestation. Villagers also continued to collect wild plants and fruits. Besides household use of these items, the local commercial cooperative provided channels through which nontimber forest products could be marketed. The village's young people reported having collected quotas of items such as rosehips as part of activities of the socialist-era youth organization (pioneers). The cooperative also bought wild mushrooms, and for a time a processing facility for them existed in the village. The intensity of collection may have been lower than in the postsocialist period, however, given the less difficult economic conditions and greater likelihood of having regular paid employment.

Rural Industries

During the socialist period, villagers also had increasing access to nonagricultural employment opportunities. Some people left for the city and sought employment in the country's growing industrial sector. Industrial development also came to the village. Beginning in the 1960s, the Bulgarian government located small production and assembly workshops in many rural villages. This was done to create an incentive for young people to stay in villages by providing them with year-round nonagricultural employment. These government efforts attempted to stop rural depopulation and the associated labor shortages that accompanied the socialist development strategy, thus reducing pressure on the limited urban housing supply (Creed 1995a). In other words, the socialist development strategy of pushing rural labor out of agriculture and into urban-based industries had been too successful, and these rural enterprises were seen as a means to stem the flow of population from countryside to city.

The first of two such facilities in Zaburdo was a sewing workshop that opened in 1973. It was initially tied to a small enterprise in a nearby village and subsequently became part of a larger sewing and textile enterprise in Smolyan. Already-cut cloth was assembled into lighter-weight men's and women's clothing, some of it for export, and the workshop also did custom sewing for individual customers. Except for managers and technicians, the sewing workshop employed women exclusively, and at its peak, it had 60 workers in two shifts. Workers at this enterprise drew together into a tight-knit social group that was evident in other aspects of village life. The women went on employer-organized excursions together, gave blood together, and participated in a folk-singing group associated with the workshop. Workshop employees were also involved in agriculture through their service as brigades on the cooperative farm during peak labor seasons. This was initially an infrequent occurrence, but as the number of workers in the cooperative farm declined over time, their participation in the work of the farm increased. Each shift at the workshop served as an independent agricultural brigade and later made up the lost work time at the workshop. In this fashion, the workshop employees earned additional income, and the cooperative farm supplemented its shrinking workforce. Workshop brigades similarly participated as production units in the *akord* system in the final three years of the cooperative farm. The village sewing workshop closed in 1989, and the abandoned building sits across the road from the village health service.[20]

A second assembly workshop opened in the village in 1982, making lighting fixtures. As one of four or five workshops affiliated with a complex-element factory in Chepelare, the workshop's workers assembled parts from the factory into lighting and electrical components for both Bulgarian and export markets. It had a smaller workforce than the sewing workshop—two brigades of 15 each, with women outnumbering men. Like the sewing workshop employees, this workshop's brigades served as production units in the *akord* system. It closed about 1990, and the facility no longer exists. Zaburdo residents who used to work at this workshop periodically appear in this work's later chapters, illustrating some of the career changes that they have undergone in the postsocialist period. The next two chapters focus on what has happened to natural resource use and the village economy following the collapse of Bulgaria's communist regime in 1989–1990.

3

Postsocialist Strategies of Mountain Agriculture

Before the sun rises on summer mornings, Zaburdo residents stir to life. After feeding their livestock and milking the cows, villagers prepare for a day of work. They line up for fresh bread at the bakery, perhaps pick up last-minute supplies at the store, pack lunches in colorful woolen bags, collect tools, and saddle pack animals. Although some people go to nonagricultural jobs after taking their milk to the buying point and sending their cows into the street to join the neighborhood herd, most head out to cultivate potatoes, harvest hay, collect wild mushrooms, or engage in other activities directly dependent on the environment. Some leave on foot, others riding or leading a mule or donkey, and during peak season there can be a long lineup of large Russian farm trucks on dirt roads leading to popular agricultural sites. The traffic jam can be particularly bad if the trucks become stuck behind a herd of a hundred or more cows.

Zaburdo is left nearly empty, except for older people who are no longer able to work and some children who are left to their own devices or under the watchful eye of a grandparent or great-grandparent. The littlest ones may be sent to the kindergarten, and many youngsters accompany their parents to the fields. Since the settlement is nearly abandoned at midday, stores, coffee bars, and the bakery are open for a few hours in the morning and then reopen again in the early evening.

In late afternoon, the pattern reverses, and the main street is again full of people and activity. Mushroom collectors straggle home to sell their bags of fungi; pack animals are led into the village with loads of hay, potatoes, or firewood; and the trucks return full of goods, workers, or both. The few shepherds bring their flocks of sheep back to their barns. Twelve hours after they left, the cows come home to waiting villagers, who again milk and feed them. After dinner and perhaps an hour or two watching the evening's offering on one of two state-run television channels, residents turn in, only to rise again early the next morning and repeat the pattern.

Following the previous chapter's historical discussion, this chapter turns to present-day resource use by describing the agropastoral activities of village residents along with other ways they exploit the mountain landscape. After discussing mountain agroecology and the land restitution process, this chapter details farming strategies, livestock production, and the collection of firewood and various "wild" resources that are found in forests and meadows surrounding the village. The concluding section engages with discussions of changing property relations and considers the implications of the land restitution for village residents and their agricultural production.

Mountain Agroecology

Particularly relevant ecological characteristics of mountain settings for agropastoral and other livelihood activities include vertical biotic zonation, irregular biotic distribution due to local variations in rainfall and daily sunlight exposure, slope, elevation, soil characteristics, and rugged terrain (Cole and Wolf 1974:122; Rhoades and Thompson 1975:543). As shown in Table 3, nearly 70 percent of the territory associated with Zaburdo is forested and about 9 percent is pastureland, leaving a relatively limited area— 9,263 decares or about 21 percent of the territory—for agricultural production (unpublished data, Chepelare Municipal Land Commission, 1997). These agricultural fields and hay meadows range in elevation from 900 to 1,700 meters above sea level, which limits the crops that can be successfully produced. Growing seasons shorten and temperatures decrease as altitude increases, which can also lower productivity (Cole and Wolf 1974:121). The mountain environment can affect pastoral activities as well. In

Table 3. Classification of land use within the village territory.

Permanent use	Decares	Percent of total
Agricultural Fields	5,922.3	13.3%
Hay Meadows	3,340.7	7.5
Pastures	3,902.5	8.7
Coniferous Forests	26,163.9	58.6
Young Forests*	4,371.1	9.8
Other**	947.3	2.1
Total	44,647.8	100.0

Source: Unpublished data, Chepelare Municipal Land Commission, January 1997.
* *Gorski traini nasazhdeniya* = forestry plantations. Land within the agricultural fund that has turned into forests in the years since collectivization.
** Includes roads, the village, agricultural facilities, the graveyard, archeological monuments, water sources, and sporting facilities.

such a setting, grazing is often seasonal, and people raising livestock must either find alternative winter pastures or accumulate resources to feed and house animals through the winter.

Soils in the area are relatively poor, further affecting agricultural possibilities. The parent rocks of soils in the village territory are predominantly of igneous origin—granites, granodiorites, schists, and rhyolites. The main soil types resulting from the breakdown of these rocks are dystric cambisols and dystric leptosols. These are primarily brown forest soils of medium to coarse texture and ranging from slightly to very stony. According to the categories of Bulgarian soil scientists, based on the suitability of soils and climates in the region for producing various crops, soils of village lands fall into the worst categories from an agronomic standpoint, that is, poor quality and suitable for only a limited number of uses—potatoes, pastures, and hay meadows (Soil Characteristics and Soil Map 1973). These are generally the uses to which they are put.

The central Rhodope climate is classified as transitional continental, with the term "transitional" referring to the influence of the Mediterranean, particularly in southern and eastern parts of the range. As with other mountainous areas, the complex topography results in local variations in both temperatures and precipitation (Danchev 1998; Knight and Staneva 1996; Perry 1995:116). Data from the nearby municipal center of Chepelare provide a long-term picture of climatic conditions in the region.[1] As illustrated in Figure 3, winters are cold, and summers are mild. On av-

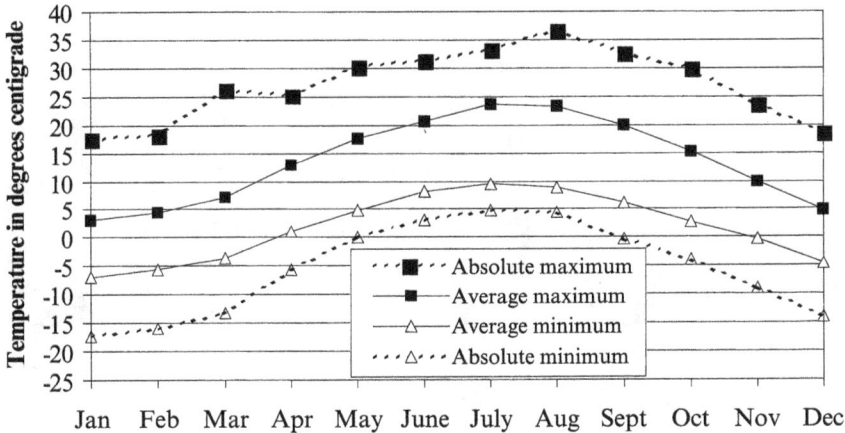

Source: Unpublished data. National Institute of Meteorology and Hydrology, Bulgarian Academy of Sciences, 1998

Figure 3. Monthly temperature in Chepelare, Bulgaria, 1921–1995

erage, minimum temperatures below freezing are seen between November and March, although they can occur as early as September and as late as April or May. Maximum summer temperatures average in the low 20s Centigrade, although they occasionally reach the mid-30s Centigrade.

The region is relatively well watered by rain and snow in average years, although periodic droughts can affect crop yields and domestic water supplies.[2] Precipitation falls year-round, with the largest amounts typically occurring in May, June, and July and the smallest amounts in March, August, and September (see Fig. 4). The long-term (1921 to 1995) average annual precipitation for Chepelare is 823 millimeters, compared to an average annual precipitation of 680 millimeters for the country (Knight and Staneva 1996:347) and 540 millimeters a year for the city of Plovdiv, located in the lowlands to the north (Velev 1996:368). These averages hide considerable year-to-year variation, however. Over the 1975 to 1995 period, average annual precipitation at Chepelare ranged from 586 to 953 millimeters (standard deviation [S.D.] 106 millimeters). Similar variation is evident in the monthly data: average rainfall in June was 94 millimeters, with a range of 28 to 230 millimeters (S.D. 52 millimeters).

These climatic conditions place some constraints on agricultural activities and lend a distinct seasonality to village life. Agriculture is almost exclusively rain fed and thus dependent upon

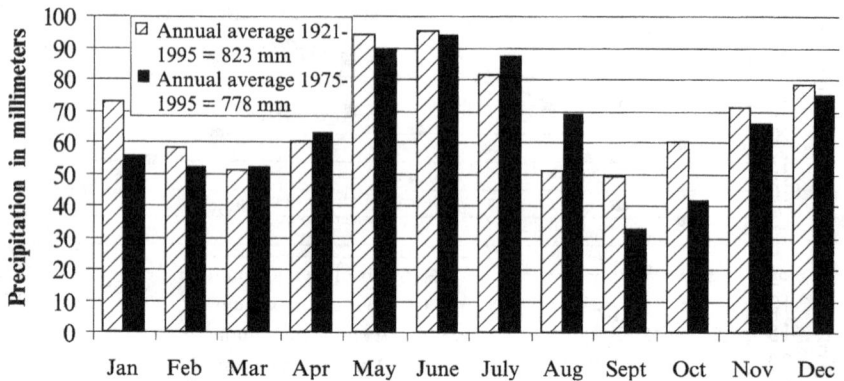

Source: Unpublished data. National Institute of Meteorology and Hydrology, Bulgarian Academy of Sciences, 1998

Figure 4. Average monthly precipitation in Chepelare, Bulgaria, 1921–1995

seasonal precipitation. Lack of rain is sometimes cited as the cause for poor potato harvests. Rain between the cutting and gathering of hay can cause spoilage, and snow can knock down uncut hay, making it impossible to harvest. Temperature conditions similarly place limits on the length of the agropastoral season. Only one potato crop is grown each year, although hay is sometimes harvested a second time. Extensive field preparation such as plowing and planting cannot take place until the snow is gone from a given field, and the pattern of cultivation moves uphill as the snow melts in the spring. Similarly, large-scale grazing cannot occur until the snow has melted and the grass has grown enough to provide adequate food for the animals. Meanwhile, farmers must produce or procure food to feed their cows, sheep, and other livestock throughout the winter. More generally, the combination of altitude and temperature, along with soil characteristics, limits somewhat the crops that can be grown successfully. For example, corn, grapes, and tobacco are not grown, and success with tomatoes is said to be poor. Another temperature concern is spring or fall frosts, which can cause crop damage. The frequency with which this happens is unclear; however, a frost in June 1997 damaged the bean crop of some Zaburdo residents. (The potatoes seemed less affected.) Very low winter temperatures can also threaten potatoes stored for food or seed if storage facilities are not adequate to keep the produce from freezing. Issues of seasonality are further discussed in chapter 4. But first I examine how villagers make use of local natural resources, beginning with the restitution of agricultural land and the dismantling of socialist-era production structures.

Restoration of Private Agricultural Land

Passage of the Law on the Ownership and Use of Agricultural Lands (State Gazette No. 17, 1 March 1991) was an early act of Bulgaria's first postsocialist parliament in February 1991. The law is designed to recreate ownership patterns that existed prior to collectivization with little regard for the viability of the ownership structure or for the people who had been working the land in the intervening years. Land is returned to its precollectivization owners or their heirs, and the preferred means for doing so is according to "old real" boundaries, essentially those that existed before collectivization. In most parts of the country, achieving this ideal was anticipated to be difficult due to intervening developments that have obliterated the landmarks used to identify the boundaries between plots that are often quite small.[3] The expected exception to this was mountain areas, where old boundaries were thought to be more readily identifiable (see Begg and Meurs 1998; Creed 1998; Strong et al. 1996). Municipally based land commissions implemented the restitution, and the measuring and mapping activities were contracted out to private firms.

The process of restoring agricultural land in the village of Zaburdo began in 1991 and was largely completed in 1994 according to land commission officials. With the restitution, about 20 percent of the village territory was transferred to private ownership. The relatively early completion date reflects the mountain setting in which it was easiest to locate the old real boundaries. Restoring land in this fashion avoided the more complicated and time-consuming process of allocating land on another basis or otherwise providing compensation for the land (Creed 1998; Strong et al. 1996:54). As well, Zaburdo was the first of ten settlements in the Chepelare Municipality for which this restoration was completed, reportedly because of the enthusiasm and efficiency of the firm retained to do the technical work.

This restitution process has not been without problems,

however, and while it may be complete in the eyes of the land
commission, the process of sorting out and restoring ownership
to specific individuals is ongoing. First, a lack of ownership documents, in part due to the burning of the village archive by a
group of communist partisans in 1944, meant that in many cases
the restitution was based on testimonials about where the precollectivization holdings had been, rather than on paper deeds. This
conceivably allowed some people to claim more land than they
were actually entitled to. Meanwhile, others did not necessarily
claim all of their land due to concerns about eventual property
taxes or lack of belief that the land would actually be restored after 50 years of cooperative production. Second, land was restored
in the names of its owners before creation of the cooperative farm
in the late 1950s, and if this owner died in the intervening years,
division of land among the heirs is their responsibility. This sometimes was and still is a source of considerable contention between
relatives.[4] One consequence is that not all families have gone
through the legal process of getting deeds in the names of individual heirs. Finally, several years after the restitution was finalized on paper, conflict emerged (it had been brewing for some
time) between villagers and land commission officials, because
the official paper maps of the land commission do not match the
mental maps that the villagers had of their former holdings. As
long as villagers simply use "their land" in a way agreed upon by
the landowners in a given location (even if this does not match
their deeds and other land ownership documents), this is not necessarily problematic. The problem emerges in situations such as
land sales, where the official paper maps come into play, particularly when a lack of correspondence between the two sets of
"maps" is found. These situations were seen to be so pervasive
that many villagers signed a letter that was sent to the agriculture
ministry in 2001, requesting that the land division plan be redone.
Their request was denied on procedural grounds. The problem is
anticipated to become more severe when older residents, who
know the old land locations, are no longer around, and there is reliance upon paper documents that don't match the ways in which
land is being used.

   The land restitution has left most village households or members thereof as landowners, yet total holdings are typically small
and fragmented. Self-reported data from my survey of one-half of

Table 4. Distribution of landholdings.

Holding Size	Number of households	Percent of households	Land area (in decares)	Percent of total area
Landless	2	1.7%	0	0.0%
Less than 5 decares	9	7.7	31	1.5
5 to 9.9 decares	20	17.1	150	7.4
10 to 14.9 decares	25	21.4	310	15.3
15 to 19.9 decares	22	18.8	351	17.3
20 to 29.9 decares	21	17.9	477	23.6
30 decares or more	18	15.4	703	34.8
Total	117	100.0%	2,022	100.0%

Source: Author's sample survey of 120 village households in 1997. Data are missing for three cases.

the village's households indicates average landholdings of about 17 decares, with a maximum holding of 92 decares, and only two households reported owning no land. As shown in Table 4, the distribution of land is relatively equitable, and even those holdings that are large by village standards are modest. Nearly 85 percent of surveyed households owned less than 30 decares of land, and these holdings represented 65 percent of the land owned by surveyed households. (In other words, the top 15 percent of surveyed households in terms of land ownership owned only about 35 percent of the land.) Village landholdings are found in named locations, and the names of these locations often refer to physical features such as the marsh, the lake, and below the rock; to cultural features such as the old village, the church, and the graveyard; and to specific families such as Slivovi's cabin or Marinovi's meadow. Reflecting the pattern of land fragmentation and dispersion seen in other parts of Bulgaria, each surveyed household owned land scattered in an average of 5.5 different named locations. Some households' holdings were concentrated in one or two locations, while one household—a large, multigenerational one—owned parcels in 14 locations.

The cooperative farm in Zaburdo was dismantled in the early 1990s in a process parallel to, although separate from, the land restitution. The liquidation process had important implications for agricultural infrastructure and resources. A village-based commission was responsible for disposing of the farm's buildings, machines, and animals, along with operating the farm until the liquidation was finished. Villagers had the first option for purchase in the livestock auctions, and some livestock remained in Zaburdo. Machines and buildings were auctioned off to the highest bidders, however, with no preference given to local residents. Conversations with the commission's former chairperson suggest that most machinery from the farm was sold to outsiders, including all the hay mowers and potato sorters, half the farm trucks, and all but one tractor. The latter was sold to the village administration before the auction for the benefit of residents. Two high-value fish farms and a few buildings were purchased by outside buyers, while more buildings were sold to villagers or groups thereof. The sales took place in 1992 and 1993, and the regional court finalized the liquidation in 1995. Some proceeds from these asset sales were distributed to former cooperative farm members based on the amount of their land and other assets used by the farm and the number of days they had worked on it. Thus, villagers were left with some financial compensation from the liquidation plus a few trucks, animals, and buildings, but other productive resources ended up in the hands of outsiders and are thus not available for use by villagers in their postsocialist production.

Postsocialist Patterns of Agricultural Production

With this liquidation of the farm and restitution of agricultural land, villagers returned to private farming in the early 1990s. Outside employment opportunities are limited, in part due to the village's location at the proverbial end of the road and the closure of socialist-era assembly workshops. Consequently residents are substantially involved in agricultural activities, including the continued production of potatoes, hay, and livestock. Indeed, those surveyed households that reported recently increasing the amount of land cultivated explained that this had been done in response to the unemployment of a household member and consequent need to generate more income. All but 2 of 120 village households surveyed were involved in agricultural production in 1997. In addition, some individuals who currently live elsewhere but own property in Zaburdo as a result of the restitution also cultivate village land. The village lacks a postsocialist agricultural-production cooperative, and agriculture is organized on a household basis.[5] This production typically takes place on the numerous small plots scattered across the landscape using the hand tools and animal traction that characterized the precollectivization period.

Farm size in Zaburdo largely parallels the results of the land restitution process. According to my survey data, the total area of land farmed by individual households ranged from a few tenths of a decare to 108 decares, with an average of 15 decares. Virtually all households farm land belonging to someone in the household. In addition, 31 percent reported using land owned by someone outside the household—in about half the cases, a relative—in amounts ranging up to 100 decares. (A few households with significant livestock holdings rent large areas of meadows to produce enough hay to feed their animals, but rented areas were usually more modest.) Conversely, 27 percent of those surveyed reported that some of their land—up to 35 decares—is rented out or otherwise cultivated by people outside the household. Table 5 lists the conditions for nonhousehold use of agricultural land. In

Table 5. Conditions for nonhousehold use of agricultural land.

	Conditions under which nonhousehold land is used		Conditions under which household land is rented out or otherwise worked by people outside the household	
	Number	*Percent*	*Number*	*Percent*
No conditions	1	2.7%	7	22.6%
As part of inheritance	2	5.4	11	35.5
For labor	4	10.8	7	22.6
For money	22	59.5	1	3.2
For produce (milk, potatoes)	12	32.4	5	16.1
For firewood	0	0.0	1	3.2
Total respondents	37		31	

Source: Author's sample survey of 120 village households in 1997. Responses shown are from households that worked land belonging to someone outside the household and households that rented out or otherwise allowed people outside the household to use their land.

Notes: Totals add up to more than 100 percent because some respondents reported a combination of conditions (e.g., produce plus money). Lack of correspondence between the two sets of data regarding inheritance may stem from households receiving land through inheritance not reporting it as nonhousehold land.

some cases, this use is without condition or payment, as when living parents give land to their children to use as part of their inheritance. In other cases, the lending or renting household receives money, produce, labor, firewood, or a combination thereof in exchange for use of its land. Thus, these arrangements indicate continuity with the presocialist pattern of smallholder production by landowners, and rental arrangements are not always monetized.

An important characteristic of agricultural lands is their distance from the settlement. While some fields are located within a few minutes walk, others may require walking for an hour or more. Often the closest fields are the smallest and have been divided the most, while those farther away can be larger. One reason given for not using some household properties was their distance from the village. This, in turn, is related to the problem of transportation—that is, transport of workers, seed, tools, and fertilizer to the fields, and transport of produce and workers back home. One technological development in the postsocialist production system of some households—in comparison to precollectivization practices—is the use of motorized transport and vehicle roads. Yet, the lack of a vehicle or the money or connections to rent or borrow a truck for a few days can make cultivation of distant parcels a chore. Household survey data indicate that 19 percent of village households own a truck that can be used for these purposes; 21 percent own a car, which may or may not be used for farming tasks; and about 10 percent own both types of vehicles. This leaves 70 percent of village households that must rent, borrow, or do without, and having a car or truck does not get around the need for cash to buy gasoline. It is not surprising that transportation issues, such as the high costs of gasoline or transportation services, were the second most frequently identified problem for household crop production (see Table 6). This sometimes has unanticipated consequences for agricultural production and resource use. One older widow, for example, explained that she had to rely on more expensive commercial fertilizer, rather than using her cow's manure, because the bags of dry fertilizer were easier to transport. The manure, which she identifies as a valuable fertilizer, is dumped in the village stream instead.

A change in settlement patterns over the last century is relevant here. Although most families had houses in the village before col-

Table 6. Significant problems in household crop production.

Problem	Number of households	Percent of households involved in farming
High cost of artificial fertilizers, pesticides, and/or herbicides	53	44.9%
High cost of transportation and/or gasoline	21	17.8
Colorado potato beetles	12	10.2
Health or age of household members	9	7.6
Lack of cash with which to purchase inputs	7	5.9
Lack of availability of inputs in the village	5	4.2
High cost of plowing and/or tractor services	5	4.2
Problems with labor organization	5	4.2
Low price received for production	2	1.7
Potato blight	2	1.7
Frost	1	0.8
No problems identified	50	42.4

Source: Author's sample survey of 120 village households in 1997. Percentages are out of 118 households involved in farming. Figures add up to more than 100 percent because multiple responses were recorded.

lectivization, many also maintained small buildings (*kolibi*, sing. *koliba*) for their animals and themselves near their larger land-holdings.[6] These two-storied stone and wood structures with stone slab roofs provided shelter for animals on the ground floor and hay storage and a small living space on the upper floor. Animals were kept in these *kolibi* near the hay meadows, thus limiting the need for long-distance transport of hay. Their manure was more easily transported to adjoining fields for use as fertilizer, again cutting down on transportation efforts. Household members often spent much of the summer and sometimes the winter as well at these locations, thus dealing with the problem of transporting agricultural labor from the village to the fields. In the postsocialist period, in contrast, a few people use outlying facilities for livestock production, but most often they maintain a residence in the village and travel daily between the village and the *koliba*. Although one encounters these buildings when traveling across the landscape, they were not necessarily maintained during the socialist period and sometimes are in poor condition. Another potential barrier to the use of *kolibi* is that they are not necessarily accessible by vehicle road, since roads were not an issue when they were built.

Virtually all village households raise potatoes. This staple food crop, and the only one covering a significant area, was also the primary crop of the cooperative farm. As the snow melts in the spring, villagers begin plowing lower-elevation fields, sometimes mixing in livestock manure or commercial fertilizer in the process, and planting them by hand with small potatoes that have been saved from the previous year's harvest for seed. The plowing and planting move progressively uphill as the season advances and snow disappears from the surrounding peaks. Potato fields are cultivated twice to control weed growth and to mound soil around the plants. The first cultivation takes place with a simple wooden-handled hoe and may involve a second application of fertilizer. An animal-drawn plow is sometimes employed for the second cultivation if this resource is available and the plants have not grown up too much. The final major activity in potato production is harvest in the fall. A plow is often drawn through the fields to dig up the potatoes, which are collected by hand and put into gunnysacks for transportation by truck or animal to the village. Potato production is a sequential process, involving multiple parcels of land scattered across the landscape. Each major activity on a plot is typically finished in one or a few days before moving on to another parcel.

In the postsocialist period, most fields are planted annually and without rotation of crops. Fields are not obviously terraced, although many plots are on a slope due to the mountainous terrain. The steepest land is generally used for hay rather than potatoes, and even so erosion damage sometimes occurs when heavy rain falls on steeper potato fields soon after planting, washing out the seed potatoes. Surveyed household reported planting an average of 3.7 decares with potatoes, although the area planted ranged from a few *ara* (a few hundred square meters) to 30 decares. Most farmers say yields range from 2,000 to 4,000 kilograms per decare; however, my rough calculations for a few households show average yields at the low end of this range, and regional statistics similarly list average yields of 1,000 to 1,600 kilograms per decare (Smolyan Territorial Statistical Bureau 1997:13). In the absence of historical yield data for Zaburdo, the regional data show no patterned changes in yield per decare or area planted with potatoes between the 1980s and 1990s. Total areas planted have remained steady, while occasional low-yield years are likely due to rainfall deficits (Smolyan Territorial Statistical Bureau 1995:53–54, 1997:86, 88). The potato varieties are a combination of those grown on the cooperative farm with periodic purchases of higher-yielding elite seeds. Nearly all surveyed households produce their own seed potatoes, but some farmers say that it is good to periodically obtain fresh seeds. About 5 percent of surveyed households had recently purchased seed potatoes.

Potatoes are a subsistence crop as well as a cash one. They may be boiled or fried, added to soups or stews, or used to fill a typically Bulgarian pastry known as *banitsa*.[7] Potatoes are neither the only nor even the main food served in most village households, however. Instead they are one among several staple starches. This may be a legacy of the crop diversity prior to World War II and the preparation of a variety of grains and legumes during the socialist era—except that during socialist times, the products were purchased with cash earned on the cooperative farm or other jobs instead of being grown by households, as in private times. Besides eating potatoes themselves, villagers boil the smallest potatoes and those damaged in harvest, eaten by mice, or spoiled by blight and then feed them to their livestock. They also save part of each year's crop to use as seed the next year.

Potatoes are also an important source of cash income and exchange value. Nearly 80 percent of surveyed households sell potatoes, barter them, or both. Villagers typically sell their potatoes in the fall after harvest, although a few wait until the winter or spring in hopes of getting a better price. The latter strategy is dependent on weatherproof storage space plus the economic ability to wait several months for the income. (Plus, if one waits too long, the lowland spring crop is harvested and the previous year's potatoes have little value.) Most households sell their potatoes in the village to people who come looking for them, exchanging them for money and sometimes goods. About one in six of the surveyed households transport the potatoes to market elsewhere and sell them retail, wholesale, or to institutional customers such as hospital or factory cafeterias. Although one may get a better price in this fashion, access to transport is required for the bulky goods, plus the time and desire to look for buyers.

Marketing is a critical issue. Although most residents grew

potatoes through the *akord* system or on private plots during the socialist period, they had a ready market for any excess produce through the cooperative farm. While the socialist distribution system has fallen apart, it has not necessarily been replaced by a producer-friendly alternative. As Verdery (1998, 1999) points out, access to agricultural land is a necessary but not sufficient condition for production. Production on the restored land and its profitability can be affected by external conditions such as the availability of the means for cultivation (e.g., implements and draft animals) and the ability to dispose profitably of the product. I will return to this issue later.

Along with the staple potato, case-study villagers may grow vegetables and herbs in small plots or at the edges of their potato fields (see Table 7). The amount of land planted with these other vegetables is usually small—a few tens of square meters to perhaps half a decare at most. Except possibly for plowing, all activities are typically done by hand, and garden care does not require the complex labor organization and scheduling seen in potato production. Although greenhouses and cold frames were observed in some Rhodope communities, I saw neither method for extending the growing season or crops grown used in Zaburdo. The average household produced about three different herbs and vegetables, besides potatoes, and some households had as many as nine or ten varieties in their gardens. These plants were raised from purchased seeds (e.g., carrots and spinach), from seeds produced by an individual in the household or a neighbor (e.g., beans), or from produce starts, either purchased in the market (e.g., tomatoes) or raised by a neighbor (e.g., kohlrabi).

In most cases, vegetables are grown in limited quantities, but they contribute to the dietary diversity of the people who grow them.[8] Beans are the most prevalent and often cover the largest area of garden plots. Several different varieties are eaten, either fresh or canned as green beans; harvested when mature, dried, and shelled for use in soups, stews, or salads; or dried green (shell and all) for later rehydration and preparation in stew-like dishes. Some bean varieties, such as the large and colorful Smilyan bean, are identified as typical regional crops, as is kohlrabi, which is often referred to locally as "our cabbage" and was grown by more than one-half of surveyed households.

Table 7. Diversity in vegetable and herb production.

Popular vegetables and herbs	Number of households	Percent of sample
Potatoes	117	97.5%
Beans	104	86.7
Onions	98	81.7
Kohlrabi	63	52.5
Carrots	30	25.0
Summer squash	18	15.0
Parsley	16	13.3

Source: Author's sample survey of 120 village households in 1997.

Note: Fewer than 15 percent of surveyed households grow the following: mint (*djodjen*), lettuce, white cabbage, celery, garlic, pumpkins, tomatoes, dill, leeks, Jerusalem artichokes, sugar beets, dock, radishes, summer savory, spinach, mountain spinach, peppers, and cucumbers.

These vegetables are grown primarily for household consumption, although some informal exchange occurs among villagers and between villagers and their families or friends in town. In the latter case, city-dwelling children or friends might be sent home with whatever vegetable was ready at the time of their visit, or with jars of produce canned for the winter. Smollett (1989) refers to the latter as "the economy of jars" and uses it as an example of the continuing importance of kinship relations in Bulgaria under socialism. Creed (1998) similarly points to the role of rural–urban connections in provisioning urban households as food distribution systems have broken down in the postsocialist period. I return to these issues of exchange networks in the next chapter.

Hay production straddles the border between crop production and animal husbandry. Like potatoes, it requires land, labor, and sometimes fertilizer, but it is also intimately tied to animal husbandry because the reason for producing hay is to feed ones' cows, sheep, mules, and so on. Hay is grown on privately owned meadows[9] and typically requires only one major period of labor for harvest in late summer, although it is sometimes cut a second time in the fall and meadows may be fertilized. Of surveyed households, 89 percent harvest hay—essentially the same number as those raising livestock—and one household with extensive meadows reported hay sales. Surveyed households harvested hay from an average of 11.6 decares of meadows, and the maximum area harvested by an individual household was 100 decares. Factoring out three households with significant livestock holdings, who consequently each harvest 50-plus decares of hay, the average area drops to about 10 decares.

Strategies of Agricultural Production

Postsocialist agriculture combines the "modern" methods used on the cooperative farm with the lower-technology practices that characterized precollectivization agriculture. Potato production takes place using simple tools, like hoes and plows, and animal traction. No individual villager owned a tractor in 1997, although by 2002 a few better-off households had purchased small ones suitable for the small parcels of the mountain landscape. In addition, the local administration operates a tractor obtained in the cooperative farm liquidation, and villagers can hire its services to plow larger fields in more level locations. Otherwise plowing involves a mule- or horse-drawn plow, and the preparation of the smallest plots may occur exclusively by hand. Plows are often used for the second potato cultivation and at harvest for digging up the plants and exposing the tubers. The first cultivation is done by hand, however, and harvesting also requires people with hoes to dig for and gather the potatoes uncovered by the plow. Thus, postsocialist potato production involves considerable manual labor combined with the use of draft animals for most households.

This return to low-technology, labor-intensive production methods of an earlier era is combined with continued reliance on the commercial fertilizers, pesticides, herbicides, and fungicides that came into widespread usage on the cooperative farm. Although most farmers use manure from their livestock as fertilizer—plowing it into their fields after the harvest and perhaps repeating the process before planting—potatoes are also often fertilized with commercial phosphorus and ammonium nitrate fertilizers at planting and sometimes during the first cultivation as well. Reflecting the costs involved, concerns about the health effects of chemical fertilizers, or both, some households report that they artificially fertilize only fields used to produce potatoes for sale, while they use only manure on the fields from which they will eat potatoes.

Village farmers also use agricultural chemicals to fight threats to the potato crop. According to a potato specialist at the agricultural research station in Smolyan, the most common problem for potato production in the region is potato late blight, and another common pest is the Colorado potato beetle. Some villagers suffered losses from blight in 1997, some had fields affected by blight but without significant crop damage, and others reported no such problems. Only two surveyed households identified blight as a significant problem in their agricultural production, however, and few villagers were said to apply fungicides to combat blight. In contrast, voracious potato beetles and what to do about them are frequent topics of discussion, and I occasionally met people on their way to spray their fields with pesticides using backpack sprayers. About 10 percent of surveyed households identified this beetle as a significant problem for their household's agriculture, and it was the third most commonly identified problem (see Table 6). Besides spraying pesticides in response to the appearance of this shiny black-and-yellow pest, some farmers also remove the beetles from plants by hand. Finally, herbicides are frequently sprayed on potato fields shortly after planting to reduce or prevent weed growth. As shown in Table 6, the most significant problem identified by farming households was the high cost of agricultural inputs such as pesticides, herbicides, and fertilizers.[10] Following collapse of the socialist system, fertilizer and pesticide prices adjusted to world levels, while the prices that farmers receive for agricultural produce have been slower to change. One reason for the latter is declining demand for food and changes in consumption patterns (i.e., substitution of lower-cost bread for more expensive meat products) in response to price inflation (Begg and Meurs 1998:260, 263). Other problems associated with using commercial agricultural chemicals are their lack of availability in the village and the need for cash to buy them. This continued use, despite their high cost and lack of availability in the village, may reflect a commercial orientation to this production.

A similar pattern of old and new practices and technologies is seen in hay production. Some villagers manually clean hay meadows of undesirable vegetation, such as rose bushes and small trees, and they occasionally employ fire in the process. Hay is cut and then collected by hand using simple scythes, wooden rakes, and wooden pitchforks. Proper meadow management, according to one villager I spoke with, involves either a second cutting of hay in the late fall or grazing livestock on the meadows after the harvest. In the latter case, the meadows are fertilized with the manure left behind by the animals. Villagers may also fertilize their hay meadows with ammonium nitrate in the spring after a rainfall. One person explained that this was particularly important for obtaining a good harvest from meadows planted in the past with a grass mixture, while production from natural herbaceous meadows was not so dependent on fertilization.

In the absence of an agricultural production cooperative, village agricultural activities are organized on a household basis. In a few cases (6 percent of surveyed households), household members (e.g., grown sons, or the sons and their wives) raised potatoes and perhaps hay independent of the household as a whole. Otherwise, production takes place with the household as a whole as the primary unit of decision making and labor organization. Decisions about crop production are made by male householders (36 percent), by male and female householders together (33 percent), or by female householders (22 percent). These figures show women's involvement in agricultural decision making, thereby cautioning against directing development or conservation-oriented interventions only at male farmers.

In most cases, all able-bodied adults participate in one way or another in the household's agricultural activities, although an individual's participation level may be influenced by involvement in wage employment, private business, or other activities. Vegetable gardening tends to be the domain of women, with the possible exception of plowing. Plowing fields, transporting produce, and other activities involving work animals, along with driving trucks or tractors and cutting hay, are generally the domain of men. Women more frequently do the hand labor of planting, cultivating, and harvesting. Men sometimes join them in these activities when not otherwise occupied, however, and women conversely may occasionally get involved in plowing or cutting hay. Children may accompany their parents or grandparents to the fields and sometimes assist with easier tasks, although this varies by family and age. Older villagers, who would have been young adults before collectivization, are often identified as having the most knowledge about private agriculture and thus as knowing things that younger generations do not. Young people may also

profess a lack of knowledge about the details of farming, defer-
ring to their elders. (A few younger villagers have studied agricultural subjects in high school or college, however, and others
have actively taken up the occupation of farmer with or without
such education.) To the extent that they are physically able to
work, older residents may also be seen as the hardest or most enthusiastic workers. They, in turn, sometimes complain about the
laziness of the village's young people and their lack of interest in
agricultural activities. Young people typically express interest in
careers that would take them outside the village—including
medicine and tourism—although some of them are also prag-
matic about their near-term involvement in agriculture given Bulgaria's current economic situation.

The organization of agricultural labor often depends on the
crop involved and the consequent labor needs. Some tasks—particularly harvesting hay and planting and harvesting potatoes—
may require more hands than are available in a given household.
Although farmers occasionally hire casual labor by the day for
cash, mutual labor exchange is much more common. More than
four-fifths of the village's households reported participating in
such labor exchange, a long-standing tradition throughout Bulgaria (Krustanova 1986).

Mutual labor exchange is coordinated at the household level.
A household collects the number of relatives, other *blizki*,[11] and
neighbors that is necessary to accomplish a given task in a particular field on a single day, it feeds them well during the work-
day, and the household later returns the "borrowed" days. Members of such groups are often relatives, but some households see
neighbors as more important. This is especially the case for those
estranged from siblings or parents due to disagreements over
property ownership or other financial matters. Households that
are particularly close sometimes discuss whose field will be
planted, cultivated, or harvested on a given day in order to coordinate activities and make sure that the work on everyone's fields
is completed. One woman described such a group to me as their
own "mini-cooperative." For less close people, one simply tries
to find the additional people needed to complete an appropriately large work-group to do a necessary task. (If necessary, workers may be hired in order to have enough people, as discussed below.) In some cases, similar work is borrowed and returned—for

example, harvesting potatoes or raking hay. In other cases, a widow might harvest potatoes or collect hay for someone in order to have a man assist with a specific and sometimes gendered activity, such as plowing or wood cutting, which she cannot do on her own and for which she cannot or does not want to pay. When not working on one of their own household's fields or that of a particularly close household, household members often scatter to help several different households or accomplish other household activities.

Although this system of mutual labor exchange takes place in opposition to monetized wage labor, considerable effort and money often go into feeding the workers a good lunch and perhaps also breakfast or snacks. The women in a household might be up before dawn to prepare the food, and the menu may include meat and dessert even if these items do not usually appear on the household's table due to cost. For the potato harvest in particular, a hot lunch is sometimes cooked on site. Often freshly harvested potatoes are boiled, although on one day the workers I accompanied were treated to a thick stew of rice and lamb. Thus, to participate in this system, one must have the labor resources to return the borrowed days, on the one hand, and the financial resources to feed the workers, on the other.[12]

Compared to the widespread pattern of mutual labor exchange, employment of casual wage laborers in agriculture is more limited in terms of both the amount of work and the number of households employing wage laborers. Twenty-eight percent of surveyed households reported hiring labor at some point during the previous year. In some cases, one or two people are hired for specialized activities such as plowing potato fields (7 percent of farming households) or cutting hay (8 percent). Plowing requires the ownership of or access to a plow and draft animals, while cutting hay is almost exclusively men's work. In other cases, one to several people are hired for activities requiring many workers, and consequently a large work group might combine paid workers with friends or relatives working on an exchange basis. The potato harvest is the most labor-intensive activity, and 18 percent of surveyed households reported hiring people for this task. Even here, the informal economy is at work. The most frequently reported means of paying workers was money (79 percent of 33 households hiring workers); however, some are compensated with firewood (3 percent) or exchange of other services (27 percent).

Members of medium-sized households, from the standpoint of landholdings and labor resources, occasionally work for wages on days when they are not needed for household activities. Such a household might also occasionally pay a worker or two to help complete some task on a particularly large field or when it otherwise lacked the resources to accomplish the task on its own. It is those households at the ends of the spectrum, however—that is, larger-scale producers and economically marginal households with few means of generating cash—for whom wage labor is most significant. The larger farmers produce potatoes for sale as well as their own use on large areas of land, and to get their crop harvested they sometimes must hire workers. Even so, some of the larger-scale producers occasionally had trouble recruiting the casual laborers necessary to accomplish a given task for cash. One man from a household with above-average landholdings complained to me that Zaburdo residents did not want to work for day wages. He attributed it to their wanting higher wages, although it may also reflect the heavy involvement of most residents in agriculture and mushroom collecting during the summer and thus their lack of time to work for others. One young woman, for example, explained that she had frequently worked for day wages in prior years but did not do so in 1997 because it was more lucrative to spend her time collecting mushrooms for sale. In another family, the adult daughters collect mushrooms, at which they are quite proficient, while their mother occasionally works as a day laborer.

At the other end of the spectrum are people from poor or female-headed households who sometimes put off their own agricultural work in order to earn needed cash or exchange their general labor for specialized services such as plowing or firewood collection. One fall day, for example, I met an older woman inspecting some bean plants that she had covered with plastic sheeting to protect them from frost. She explained that she had not picked her beans before the frost because she was busy harvesting potatoes for others so that she could earn cash and was obligated to work for several more days before she could harvest her beans. A second example illustrates the situation for a poor household and specifically how the need to generate cash affected the household's production.

The Marinovi household (Marinovi is a pseudonym, as are other names of Zaburdo residents used in this work) consists of an older couple and their adult daughter.[13] All are unemployed, although the wife is retired and received a small pension of about $16 per month in late 1997. Their situation is further complicated by limited land ownership, limited involvement in informal networks of labor exchange, and health problems. They grow potatoes, onions, beans, kohlrabi, and hay on less than 4 decares of land, and raise a few animals. Most of their fields are small, and much of the work must be done by hand. Potato production from their limited holdings is so low that the family has few potatoes to sell or barter. With limited resources for buying livestock fodder, milk production is low. Even so, the female householder periodically tries to sell milk to generate some cash. She also sometimes does agricultural day labor for other people to earn cash, and consequently she ended up harvesting her own potatoes after the snow fell. Because the husband is in poor health, they must sometimes pay for help out of their meager cash supply for someone to take his turn with the cow herd and to mow their hay. As well, cash is necessary for buying bread at the bakery and paying their electricity bill. Little if any money remains for commercial fertilizer or concentrated forage, leading to a vicious circle of low production. This example shows the constraints placed on the agricultural production of a household with limited land and labor resources and also its strategies for generating cash.

Postsocialist Patterns of Animal Husbandry

The long tradition of animal husbandry in the central Rhodope Mountains continues in the postsocialist period, albeit at a lower level and with a somewhat different composition of animal varieties than during earlier eras. In a pattern similar to that seen in agriculture, decisions about what animals to raise and what to do with them are made by male heads of households in 38 percent of the cases (out of 106 households responding to this survey question), by couples jointly or households as a whole in 36 percent of the cases, and by female heads of households in 11 percent of the cases. Nine out of ten Zaburdo households raise some combination of cows, sheep, goats, work animals, pigs, and chickens. As with agricultural land holdings, households vary in the number of animals that they own, but the range is not large. Relatively few households have no livestock and relatively few raise many animals. About 22 percent of village households keep a few chickens, primarily to produce eggs for household use, and 12 percent raise pigs for meat. While these animals are important for putting food on the table, more significant for both natural resource use and household provisioning are cows, sheep, goats, and pack animals. It is to these that the discussion now turns.

Animals taken out to pasture typically graze on lands within Zaburdo's territory. As far as I can determine, few if any pasture-lands around the village were privately owned before collectivization, and most herding took place on communal or jointly owned pastures and forests.[14] Thus, the restitution of agricultural land did not result in private ownership of pasture lands. The other relevant outcome of the restitution, however, is that an area of land nearly as large as what is owned privately is currently managed by the municipality and predominantly used as common pastures. These lands were not claimed in the agricultural land restitution, but they are also not part of the forest fund. In principle, livestock owners pay a small fee to the local forest enterprise for use of lands the enterprise controls as pasture. Finally, newly privatized agricultural land also provides some grazing resources, in that animals are sometimes taken there to graze following the hay and potato harvests. Access to privately owned meadows and fields is limited to their owners during most of the year, and one task of shepherds and other herders is to keep their charges out of other people's private parcels as the animals travel daily between the village and the pastures. After the harvest and through the winter months, more widespread grazing is allowed on the hay meadows.

Both before and during socialism, first the community and then later the cooperative farm employed a field guard (*pudar*) who was responsible for addressing livestock trespass into fields and meadows (see also Sanders 1949:10). During my fieldwork, the village administration applied for a field guard as part of an employment-creation program, indicating the continuing value placed on such a position. I rarely heard about conflicts involving animal trespass, although some of the herders and shepherds I accompanied made visible efforts to keep their livestock out of the fields of others, which were sometimes unfenced.

Beyond pasture access, animal husbandry also requires procurement of animal food. Hay production is important for feeding animals during the winter when they are not taken out to pasture, and livestock are also fed hay in the summer to supplement their grazing. Zaburdo residents rarely collect leaves and branches from pine or hazelnut trees for their animals, although this practice appeared more widespread in a neighboring village. Livestock are also fed concentrated forage, which is identified as important for milk production. Before collectivization, grain crops were produced in Zaburdo, and, during the socialist era, residents could obtain concentrated forage from the cooperative farm for their few private animals.[15] With the liquidation of the farm, there is no regular source of concentrated forage in the village, and when it is available from elsewhere, as when traders come to the village with trucks full of grain, it is reportedly expensive.

Thus, it is not surprising that the number-one problem, identified by 61 percent of interviewed households in response to a question about significant constraints on their livestock production, was the cost of concentrated forage (see Table 8). This problem, mentioned by nearly all households identifying problems, often appeared in combination with the second- and third-ranking problems. Forage is expensive and it is unavailable in the village, villagers say. On top of that, they add that they don't get paid a decent amount of money for the milk they sell. Some people reported reducing their livestock holdings in response to the high cost and lack of availability of forage, although others expressed a desire to increase animal numbers in order to generate more income through additional product sales. Households reporting no problems in their livestock production often explained this in terms of their ability to transport forage from lowland areas, thereby confirming that access to this input is problematic. Wildlife was not mentioned when householders were asked generally about production problems, and only about 11 percent of surveyed households reported that wildlife had caused problems

Table 8. Significant problems in household livestock production.

Problem	Number of households	Percent of households raising livestock
High cost of concentrated forage	65	60.7%
Unavailability of concentrated forage in the village	38	35.5
Low price paid for milk	20	18.7
High cost of insemination services	7	6.5
Problems with hay production	5	4.7
Problems with hay transportation	4	3.7
Health or age of household members	3	2.8
Lack of mill for forage in the village	1	0.9
No problems identified	41	38.3

Source: Author's sample survey of 120 village households in 1997. Percentages are out of 107 households raising livestock. Figures add up to more than 100 percent because multiple responses were recorded.

by attacking their livestock or poultry in recent years when specifically questioned about problems caused by wildlife.

Cows are the most frequently owned animals (see Table 9, which summarizes village livestock holdings in 1997), and all of the adult animals are female, reflecting their use for producing milk and calves. (Artificial insemination eliminates the need for bulls.) Male calves are raised for meat, while female calves may be raised to be milk cows or for meat. When talking about their cows, villagers are quick to point out that their animals are small but well adapted to the difficult mountain environment and suited to walking long distances in search of mountain pastures. Village cows are usually a cross between a local shorthorn breed and a Jersey variety originating from Belgium and Holland.

During the winter, cows and calves are kept in barns within or on the outskirts of the village, while virtually all cows are sent out to pasture during the day in the summer. The traditional summer herding season for cows as well as sheep extends from *Gergovden* (St. George's Day on May 6) to *Dimitrovden* (St. Dimitur's Day on October 26), although the exact dates in a given year depend on weather and pasture conditions. During my fieldwork, for example, an older villager explained that there were two "fingers" of grass—grass at the height of two finger-widths—in early June

Table 9. Summary of village livestock holdings.

	Cows	Sheep	Goats	Work Animals
Percentage of surveyed households (sHH) owning:				
—adult animals	84%	25%	3%	56%
—juvenile animals	65%	19%	1%	n.a.
Typical no. of adult animals per sHH	1 or 2	6 or fewer	2	1
Max. no. of animals per sHH:				
—adult animals	6	50	5	3
—juvenile animals	5	10	4	n.a.
Estimated total no. of animals in village (1997)	375–400	350–400	20–25	150–170
Historical comparison:				
—In 1966 (cooperative farm)	503 (258 private)	5,431 (860 private)	150 (all private)	80 mules (cooperative farm) 82 donkeys (private)
—In 1947 and 1957 (decade before collectivization)	about 300	about 3,300	86 and 151	149 and 249 mules 141 and 95 oxen 34 and 46 horses 32 and 38 donkeys
—In early 20th century		16,000	6,000	
Organization of herding labor in 1997	Neighborhood system, villagers take turns	No organized system, or regular paid herders	No organized system or regular paid herders	No organized system but not apparently an issue

Sources: Data labeled "surveyed households" or "sHH" are from the author's sample survey of 120 village households in 1997. Estimates of total number of village animals are based on extrapolation from my survey data, conversations with the village veterinarian and people who coordinate the village cow herds, and my field notes. Historical data for 1947, 1957, and 1966 are from the Smolyan regional archives (Fund 894, op. 1, a.e. 14–16). Early-20th-century data are from Primovski (1973: 313–14).

1997, when the upper neighborhood started its herding, while the previous year there had been four or five fingers of grass 15 or 20 days earlier.

Nearly all village households with cows participate in a communal system for pasturing them during the summer that originated in the early 1990s. Before collectivization as well as for the first year or two following liquidation of the cooperative farm, there were regular paid herders.[16] A situation then developed in which no one could be found to do this job, and many villagers lacked the money to pay herders. Under the circumstances, cow owners saw no choice other than to employ a system of mutual labor exchange for herding. Along with daily herding during the summer, the resulting neighborhood groups sometimes engage in related activities, such as repairing watering sites and damaged places along the herding route.

In this system, the village is divided spatially into four parts or neighborhoods, and an individual in each neighborhood is responsible for organizing the herding. For the three larger herds (100 to 140 animals each), each household is responsible for providing someone to go out with the herd, according to dates determined by lots. Every month or so the organizer goes around the neighborhood, and each participating household pulls dates out of a box to determine their herding days—typically one day in each round for each animal sent with the herd. A set rotation is used in assigning herding responsibility for the fourth, smallest herd (38 cows). Depending on the number of animals and time of year, two to four people typically go out with a herd on a given day. The people fulfilling the herding "lots," as the villagers refer to them, are usually male. The only common exception is when several family members, including a wife, a child or children, or both, go out together to fulfill the household's responsibilities for the month on a single day. An experienced person on the team decides upon the day's route based on knowledge of where the herd has gone the previous day. Through this system, cow herding is an occasional rather than an everyday activity, and labor is freed up for other tasks. While some people identify herding as arduous work involving much walking, others consider their turn with the herd to be a welcome break from their regular jobs in the local school or forest enterprise.

Despite the random assignment of herding dates, this system

includes mechanisms to accommodate other responsibilities and labor constraints. Dates can be traded as long as someone can be found who is willing to do so and has a preferred date. It is also possible to pay someone to go in your place, and each neighborhood has a set daily rate for herding (e.g., 4,000 or 5,000 leva—about $2.25 or $2.75—in 1997). About 19 percent of surveyed households participating in this system regularly pay someone to take their herding duties due to employment commitments, advanced age, or disability. Other households occasionally do so when illness or other obligations prevent them from taking their turn. For people working as paid herders, this is an occasional source of cash income (unless, of course, they are trading herding for agricultural labor).

In the fall, after the hay harvest and as agricultural labor obligations decrease, households may move into a different system. Some farmers pull their animals out of the neighborhood system early, explaining that they have fewer obligations and that there is a lot for the cows to eat in their hay meadows, while the animals sent with the neighborhood herd come home hungry. In other cases, the animals are taken out in smaller groups, often to harvested hay meadows, only after the neighborhood system ends. In either case, households may cooperate with a small number of relatives, neighbors, or both in taking out the animals, again demonstrating the continuing importance of kinship and friendship networks.

Some households continue the long-time Rhodope practice of sheep production. Overall numbers are considerably reduced from the precollectivization era, however, and the proportion of village households with sheep is significantly lower than that with cows. Four out of five surveyed households with sheep send them to pasture in the summer, and about one-quarter of surveyed households also do so in the winter when snow conditions permit. Typically the latter own larger flocks and have someone who regularly acts as a shepherd. The earlier pattern of seasonal transhumance, in which flocks were taken to graze in the lowlands for part of the winter, has disappeared. This means that hay and fodder must be procured and stored to feed the sheep during the winter. The reason that not all households send their sheep to pasture in the summer involves the organization of herding labor, and this may also influence the number of households who choose to raise sheep. That is, some may have no sheep at all due to difficulties with organizing herding labor. Zaburdo currently lacks a regular paid shepherd, and no widespread system for herding sheep has emerged in the village in the postsocialist period.[17] As with cows, shepherding is a male-dominated activity, and households with larger flocks generally have a person dedicated to shepherding. In one case, two brothers in their twenties share the labor responsibilities for their joint flock of nearly 100 sheep. In this way, each goes out with the flock every other day, freeing the alternate day for nonpastoral activities. Alternatively, households make informal arrangements with a friend or relative who has a larger flock and regular shepherd to take out their sheep. The third option is to send one's sheep to a neighboring village that has a regular paid shepherd during the summer. Finally, a few households keep their animals in the barn year-round because they lack access to a regular shepherd.

As with cows, decisions concerning herding routes for sheep are made by the shepherds. The daily route is determined according to the season, weather conditions, and where the flock has been the day before, and some shepherds have particular territories where they go most frequently. Certain sunny places are more likely to be snow free in the winter, for example, while others might have a good water source or shade supply for warm summer days. Avoiding the previous day's route was explained as being important for having an adequate supply of fresh food.

Goats are sometimes described by Bulgarians as a poor person's cow, because of the relatively low cost of their care. They are also seen by foresters as a threat to young trees due to their eating habits, and forestry officials sometimes ban them from grazing on forest lands because of concern over their impact on vegetation. Very few Zaburdo households own goats, and the numbers owned are quite limited. Two of the four surveyed households owing goats take or send their animals to pasture in summer, and the other two do not. The only regular goat herder I know of is an octogenarian, who takes out his household's animals every day, along with a few goats from other neighborhood households. In other households, goats are occasionally taken out when a household member is not otherwise occupied. As with sheep, one factor affecting the number of goats in the village is likely the lack of herders or another system for regularly taking them out to graze. (Netting [1981:28] also cites the lack of a herder as leading to a decline in goat numbers in a valley in the Swiss Alps.)

With the return to animal traction and transport following the liquidation of the cooperative farm, ownership of mules, horses, and donkeys is important for village agricultural production. (Some village households also rely on cars or trucks for some transport services; however, this requires road access to meadows or fields.) Mules are the most popular, making up about half of village work animals, reportedly because they are the most stable and best adapted to local conditions with narrow and sometimes steep mountain trails (cf. Primovski 1973:345–46). Most households own only one pack animal, although some have two and rarely three. Households with only one sometimes join forces in order to have a team for plowing. Since plowing and transport are typically male activities, it is not surprising that these animals are most often cared for by a male householder (67 percent). They are periodically allowed to graze in summer during breaks in work activities or on days when they are not needed for work, and otherwise they are fed hay and concentrated forage.

Livestock Production and the Household Economy

Zaburdo residents raise cows, sheep, and goats for milk, meat, and hair or wool, and this production is for both household self-provisioning and cash generation. Milk is perhaps the most important of these products. Yogurt, butter, cheese, and dishes prepared from them frequently appear on the kitchen table and in lunch bags. Sheep milk has higher cultural value in the village, although households more commonly have cow's milk on hand simply because cows are now more common than sheep. Milk is also sold for cash to a single buyer in the village, a private dairy owned by an outsider, and thereby provides an important source of year-round cash income (in contrast to the seasonal, lump-sum payments from potato sales). Villagers are paid monthly for their milk, and the low price they receive is a regular complaint. They can also obtain from the milk buying point butter, a feta-like white cheese, the yellow *kashkaval*, and a whey cheese similar to ricotta, and the value of these products is deducted from their milk earnings. (Some households make their own cheese and butter, although Zaburdo residents frequently mentioned obtaining these products from the dairy.) According to my household survey data, virtually all households with cows sell milk regularly, suggesting some level of commercial orientation in this production, and about one-third of sheep owners and goat owners sell milk. More than three-quarters of them sell milk frequently— four or more times a week. Villagers often describe their animal varieties as being small in stature compared to those in the lowlands, but they also emphasized that such animals give milk high in butterfat. Indeed, the butterfat content of the milk and how its composition might be related to feeding practices is frequently a topic of discussion. This topic is of more than just passing interest, because the price villagers receive for their milk depends on its butterfat content.

Villagers also raise livestock for meat, primarily for subsistence consumption. A particularly anticipated festival dish is spit-roast yearling lamb or sheep (*cheverme*), and meat is also eaten fresh, canned, made into sausage, salted and dried (*pasturma*), or frozen for future consumption. Only 4 percent of surveyed households raising cows or calves and 20 percent of those with sheep reported sales of livestock for meat. Livestock are sold year-round depending on the age of the animals and levels of demand, although lamb sales are frequently higher in the spring for Easter. (The latter requires planning about when the lambs will be born.) Larger producers sometimes transport their animals to market, but sales often take place when residents or others seek out animals to purchase in the village. One producer said that he preferred sales in the village, because the animals lose weight during transport and thus bring in less money.

Wool, hair, and animal skins are currently more the byproducts of animal husbandry than its objective. Sheep are shorn in the spring, and the resulting wool is used for clothing or textile production or sold. Goat-hair rugs are one of the traditional textile products woven in this village; however, goats are few, and thus the small amount of hair produced is likely insufficient for supplying village weavers. Animal skins are sometimes sold, often to outside traders who come looking for them, while the local consumer cooperative sometimes buys wool. Alternatively, traders occasionally come to the village, offering to trade a combination of cash (for the labor) and wool (as the raw material) for woolen blankets. About 20 percent of surveyed households sold skins or wool, although this is not a significant income source. A shepherd complained to me one day, for example, that wool prices were low, and so his wool was in the barn, where mice played among the bags.

Another byproduct of livestock production in Zaburdo is manure. Before collectivization, manure played an important role in maintaining soil fertility. On the cooperative farm its use continued in combination with commercial fertilizers. Manure is still used for this purpose in the postsocialist era, and a common sight on snow-free winter days and in early spring is men transporting bags of manure from their barns to their fields with pack animals. Other people use or hire a truck for the transport. This transport is time and labor intensive, however, and consequently a portion of this resource ends up in piles behind barns, at the end of the settlement, or in the village stream. It is replaced, instead, with more costly but more easily transported commercial fertilizers.

Hunting and Gathering on Europe's Margins

Beyond intensive use of the local landscape for agropastoralism, Zaburdo residents make extensive use of several other resources available in the surrounding territory. They gather wild greens, herbs, fruits, and nuts; collect mushrooms, snails, and firewood; and hunt local wildlife. The resulting products are used for household consumption and are also sold for cash. In some ways these activities are a continuation of presocialist and socialist-era practices, although the significance of the specific items gathered and hunted and the reasons for doing so have changed over time. Their general importance in the postsocialist period is seen in that 83 percent of surveyed households gather wild products, all heat their houses with firewood, and 9 percent have members who hunt or fish. Zaburdo is not unique in this regard. Humphrey (1998:459) discusses the importance of hunting and gathering for household subsistence in rural Russia during the postsocialist period, and Creed (1998) describes commercial exploitation of gathered products in a village in northwest Bulgaria.

Nontimber Forest Products

Nontimber forest products are collected from the forest lands and pastures surrounding Zaburdo. Much of this territory is controlled by the local forest enterprise or is under temporary management by the municipality. Access to these lands for collecting nontimber forest products is not, in my experience, strictly regulated or controlled.[18] Wild harvesting of medicinal and other plants and their trade are subject to legal restrictions and prohibitions issued by the Ministry of the Environment, designed to protect specific species. A new law on medicinal plant protection and use was passed by Bulgaria's parliament in 2000, with limits on how much is considered subsistence collection and requirements that permits be obtained for collecting commercial quantities (State Gazette No. 29, 7 April 2000). These regulations were not particularly evident at the local level, although they influence the products for which there is a market. When herbs and mushrooms are collected for sale, forestry authorities collect a tax from buyers, although rates are low and receipts are small (Lange and Mladenova 1997). Thus, while some concern exists at the national level about protecting such plant populations, on the local level there is little evidence of restrictions on the collection of specific species or access to land.

Nearly three-quarters of surveyed households collect herbs, wild greens, and rosehips. These items contribute to household subsistence and also generate cash income for some households. Rosehips, the fruit of the dog rose (*Rosa canina*), are used in tea and marmalade, and some villagers recognize them as a source of vitamin C. Herbs, along with the leaves or flowers from various plants, are used for tea, seasonings, and medicinal purposes. When visiting village homes, I was often offered a cup of "mountain tea," usually prepared from a mixture of locally gathered plants. Indeed, while the U.S. media was being inundated with stories about the potential value of Saint John's wort (*Hypericum perforatum*) as an antidepressant, I was drinking tea made from this yellow-flowered plant every day, oblivious to the attention it was receiving. Other items commonly collected by Zaburdo residents for personal use in tea or for medicinal purposes include raspberry leaves (*Rubus idaeus*), blueberry leaves (*Vaccinium myrtillus*), cowslip flowers (*Primula officinalis*), ladies bedstraw (*Galium verum*), wild thyme (*Thymus serpyllum*), wild oregano (*Origanum vulgare*), yarrow (*Achillea millefolium*), and coltsfoot (*Tussilago farfara*). Wild greens are collected from pastures and hay meadows in the spring and used in cooking. Notably, a green referred to locally as *skripalets* is used like spinach in banitsa, soup, and other dishes. Beyond this, new growth of pine trees is collected to make a thick syrup referred to as pine honey, which is eaten like jam or used to treat coughs. Blossoms from the plant of the black elderberry or black elderberry blossoms (*Sambucus nigra*) are used to make a refreshing beverage or syrup and sometimes put into tea.

Bulgarians have long exploited various plant products for medicinal and related purposes, both for personal use and as an income source, and the country is one of the world's leading exporters of plants for such uses (Lange and Mladenova 1997; Lange and Shippmann 1997; see also Hardalova et al. 1998; Kuipers 1997). Given the popularity of herbal remedies in the country—one can shop for them in special herbal pharmacies as well as obtain some of them at regular pharmacies—I expected there to be greater reliance on and knowledge of medicinal plants than I found in Zaburdo. The person who really knew about this, I was told, had moved "to town," and few people I spoke with used more than a few simple remedies. For example, tea made from coltsfoot leaves or cowslip flowers is used as a cough remedy, and herb tea generally is said to be calming for the nerves. Reliance on, and thus the importance of, herbal remedies in the village may have declined with the increased availability of low-cost or free medical care at the health service during the socialist period.

Beyond this household use, herbs, flowers, and rosehips are also a source of cash income for some villagers. Indeed, collecting these plant products may be more important today for some households as an income source than for their use in folk remedies. Collection for commercial purposes dates back at least to the socialist period in Zaburdo, when the local commercial cooperative reportedly purchased "20 to 30 tons" of rosehips each year. Village schoolchildren in the 1970s and 1980s collected quotas of items such as rosehips as members of the Pioneers, a socialist-era youth organization, and such collection continues as a way for some schools to generate income in the postsocialist period.[19] The items bought at a given place and time depend upon market conditions and, obviously, when different items are ready for harvesting. The local commercial cooperative bought yarrow, wild thyme, and coltsfoot in 1996, for example, and rosehips, black elderberry blossoms, and Saint John's wort in 1997. (It also purchased edible snails [*Helix pomatia*] in spring 1997, and for a short period both adults and children could be seen searching for the gastropods.) Private buyers also bought rosehips in 1997, and several villagers related that they had collected and sold tens of kilograms of this item. Buying points elsewhere in the region purchase other items including cowslip, wild oregano, burdock root (*Arctium lappa*), and juniper berries (*Juniperus communis*). Selling these items outside Zaburdo, however, requires knowledge about what is being purchased where and the ability to transport the products to a buying point.

The sale of wild-collected plant products does not appear to be a significant income source for most Zaburdo residents, given their substantial involvement in other activities. One day, for example, a woman explained that she would not sell the rosehips she had collected because it was not worth the effort involved. It was worth her time to collect enough to make marmalade for household consumption, however. Even so, sales may provide welcome income for economically marginal individuals or those with fewer obligations on their labor. In discussing wild herbs and medicinal plants in Greece, Forbes (1997) suggests that collecting herbs for personal use is considered an enjoyable activity, while resorting to the sale of such items is a sign of great poverty and most frequently engaged in by poor widows (see also Clark 1997). While I was not aware of any particular stigma on herb sales in Zaburdo, it did appear that it was a relatively minor income source.

About 70 percent of surveyed households reported collecting locally available wild fruits and nuts including blueberries, raspberries, apples, plums, blackberries, cornel cherries, cherries, strawberries, and hazelnuts. Although sales of blueberries, for example, are seen in other parts of the Rhodope, fruit or nut sales were not evident in Zaburdo. Instead, the fruits are made into jam, compote, or syrup for home consumption, particularly during the winter when fresh products are rare. Time constraints affect the extent to which such products are exploited, even for household use. This is particularly true when the fruit ripens during the busy agricultural season. Yet women sometimes set aside a day or two for berry picking to ensure that they have jam or syrup for the winter. Of course, fruit quantities in a given year depend on weather conditions in this mountain setting. During my 1997 fieldwork many wild fruit trees and bushes had little if any fruit, perhaps due to a late spring frost, while there seemed to be bumper crops of wild plums and raspberries in 2000, when many other items (hay, mushrooms, potatoes) were scarce due to the heat and drought.

In contrast to herbs and fruits, collecting wild mushrooms in local forests during summer months is a significant source of income for many households. Of surveyed households, nearly two-thirds report that one or more members—an average of two household members—collect mushrooms. Mushroom collection is a major activity during the collecting season, which lasts for two to three months depending on weather conditions.[20] According to my household survey data and interviews with mushroom collectors, most collectors go out at least weekly, and more than one-half of them do so three or more times a week. Indeed, during the height of the season, it sometimes seems that everyone from 7 to 70 years of age is out collecting, and the woods are literally full of people. About half the mushroom collectors on a given day go out with a companion, and these groups of friends and relatives can include as many as six people. When people go collecting together, they often divide the resulting income evenly among group members. Married couples may simply put the income into the household budget without division, and sometimes each group member collects for him- or herself.

Although several edible mushroom varieties grow in the Rhodope, collection concentrates on two species. The most valuable mushroom from the standpoint of sales is the king bolete or edible boletus mushroom (*Boletus edulus; porcini* in Italian). Chanterelle mushrooms (*Cantharellus cibarius*) are also collected for sale, although their price and thus the income-earning potential is lower. All mushroom collectors interviewed had collected edible boletus mushrooms on the day of the interview, and 72 percent had also collected chanterelles. One mushroom buyer said that he could also buy slippery jack or yellow-brown boletus (*Bo-*

letus luteus) and fairy-ring mushrooms (*Marasmius oreades*). Both of these varieties are found in the region, although I never observed sales of them. Avid mushroom collectors report that other edible varieties found locally include the parasol mushroom (*Lepiota procera*), the meadow mushroom (*Agaricus campestris*), and the saffron milk-cap mushroom (*Lactarius deliciosus*).

From the standpoint of their potential contribution to the postsocialist household cash budget, mushrooms are by far the most important wild product gathered. Fifty-nine percent of surveyed households—almost all of those that collect mushrooms—reported income from mushroom sales.[21] Although some sale of wild-collected mushrooms took place before 1989, the relative importance of this activity as an income source for villagers has increased in the last decade. Most mushrooms are exported to Western Europe, with Italy identified most frequently by the village's mushroom buyers. In discussing this activity in another Bulgarian community, Creed (1998) observes that exporting wild products to European Union member states is easier than exporting commercially produced ones. In Zaburdo the chain of trade takes place through village residents—often younger entrepreneurs—acting as middlemen or -women, and there were at times half-a-dozen mushroom buying points during my summer there. (An official from the local commercial cooperative explained that it does not buy mushrooms because it cannot compete with the prices offered by private buyers.) Some local collectors also occasionally sell mushrooms at a mountain chalet about 10 kilometers south of Zaburdo or transport their mushrooms to other settlements in hopes of obtaining a higher price. The chain is also said to involve economic groups referred to locally as the "mafia," although exactly what this means is unclear, since this term is often used to refer generally to people who appear well off economically and are perhaps involved in business rather than production or wage employment (on this phenomenon, see Creed 1998; Humphrey 1999; and Wedel 1998a, 1998b). Local residents do most of the collecting around Zaburdo, but other mountain locations also see large numbers of nonresidents—the Roma minority is often mentioned—who collect mushrooms as a way to earn income during a difficult economic period.

Although much of the profit from mushroom collection is said by some villagers to go to the buyers and others further up the sales chain, collectors can still earn considerable sums compared to Bulgarian wage levels. According to data collected mid-season 1997, when prices were moderate due to the large quantities going into the market, daily per-person sales averaged 11,800 leva ($6.50), which was more than twice the typical daily wages of 5,000 leva ($2.75) from agriculture or herding cows; and one person reported selling mushrooms worth 45,900 leva ($25.50) on the day of the interview.[22] The latter sum, earned by a woman renowned for her ability to collect large quantities of mushrooms, was more than the monthly earnings of some state-sector employees at the time, and few employed villagers made more than two or three times that amount in a month.

Another way of looking at the importance of mushroom sales as an income source is to examine them in relation to other economic activities. The dedication with which the female householder described below collects mushrooms is not necessarily typical, but this case study is an example of what is possible. Sonya and Valentin Asenovi are a married couple in their thirties, and they have two school-aged children. Sonya has a regular job in the village, and Valentin has been unemployed since the closure of the village's electronics assembly workshop, where he worked. Because this couple has limited landholdings, they rent 13 decares of land for a combination of cash and produce to grow potatoes (3 decares), hay (10 decares), kohlrabi, onions, summer squash, spinach, carrots, beans, and lettuce. This household most often exchanges labor with Sonya's siblings and parents and is usually able to mobilize adequate labor to finish agricultural tasks without needing to pay cash for agricultural labor. Besides producing enough potatoes for household consumption and seeds for the next year, the Asenovi will sell a couple of tons for money and barter about half-a-ton for other goods. Excess milk from their cow is regularly sold, but deducting the costs of butter and cheese obtained from the dairy, little if any cash income is taken home. All household members collect wild fruits, snails, herbs, greens, and mushrooms, and some items are sold as well as used for household consumption. Valentin does some woodworking, and Sonya occasionally earns income from selling tufted-wool blankets and goat-hair rugs that she weaves. Through hard work in this diversity of activities, this household is able to keep food on the table, pay its utility bills, and buy some nonfood items, including a new television set in 1996.

Estimating the relative contribution of major economic activities to this household's budget illustrates the importance of wild product collection for cash generation and thus the importance for postsocialist survival. During summer 1997, Sonya earned less than 50,000 leva ($28) a month from her public sector job, and the household periodically received financial assistance related to the husband's unemployment but never amounting to more than 30,000 to 40,000 leva a month ($17 to $22). Sonya collected mushrooms after work, as well as on weekends and the occasional days that she took off. Over a two- to three-month period during the summer, she earned about 1.2 million leva ($660) from collecting mushrooms and some rosehips. This is about 20 percent more than the $550 she generated from selling 2,500 kilograms of potatoes, and represents one-quarter to one-third of the household's annual cash income. Sonya explained that this mushroom income was critically important, because a single pair of shoes for her growing children costs about the same as her monthly salary. With the mushroom money, she would buy shoes for the children. This example illustrates the relative importance of income from collecting mushrooms and other wild products for some households, as well as the diversity of activities in which this household is involved, as are many other households in the village.

Firewood for Heating and Cooking

Local forests also provide villagers with fuelwood for heating and cooking. All surveyed households heat their houses with wood, and for nearly all of them this is their primary heat source. Firewood usage averages about 12 cubic meters per year, according to self-reports on my survey, although usage varies by household size and the number of rooms being heated and thus ranged from 5 to 30 cubic meters. Many households also use wood for cooking at least occasionally, even if they own electric ranges, and some households use wood-fueled water heaters. These appliances may replace electric ones that have been unplugged in response to rising electricity prices. The later practice reflects increasing reliance on natural resources as utility prices have increased in the postsocialist period.

With this reliance on fuelwood, its procurement—almost always by men, although wives occasionally go along to help—is another important activity. In most Zaburdo households, a household member (83 percent) collects the wood, or a relative outside the household provides it (8 percent). Zaburdo households rarely buy their firewood from the local forest enterprise (7 percent) or receive it as rent for land (2 percent), and these cases typically involve elderly residents unable to collect firewood for themselves. The collection takes place in the state-controlled forests around Zaburdo.[23] For what was described as a small or symbolic fee to the local forest enterprise, villagers can collect dead and downed wood.

The timing and labor requirements associated with firewood collection often depend on the method used to transport it. People relying entirely on pack animals (42 percent) may collect wood little by little when they have time. In contrast, people using a household truck (14 percent) or a truck belonging to someone outside the household (34 percent) can procure a winter's wood supply over the course of a few postharvest days in the fall. There is a striking contrast between a group of several male relatives unloading a large Russian truck full of wood and an old grandfather leading a donkey up the road with a much smaller load of wood tied to a packsaddle. Chain saws are another useful but expensive tool in supplying a household with firewood. One man showed me a new one bought for 960 German marks (roughly $530), commenting that the monthly salary from his forestry job was 130 marks ($72). On another day I was with a villager who was pricing new chain saws at the Plovdiv Trade Fair. The Husqvarna model in which he was interested cost about $515, a considerable sum given income levels, but one that villagers sometimes try to find because of the importance of firewood for their subsistence.

The Hunt Is Up

A frequent sight early on weekend winter mornings is a group of village men leaving the settlement. Dressed in boots and camouflage clothing or hand-knit wool sweaters, they carry guns, and some lead—or are led by—dogs trained for this activity. These men are members of the local hunting society, which has about 25 members. (Although some Bulgarian women hunt, this is exclusively a male activity in Zaburdo.) These hunting parties ranged in size from half-a-dozen to more than two dozen people on days I accompanied them. Besides being an anticipated winter pastime, these hunts were a source of fresh meat for the table at a time when money was scarce for buying meat, which is relatively expensive.

Hunting is typically a group activity, and hunts begin with a meeting to organize a strategy for the outing. Older, more experienced hunters are looked to for direction about where game is mostly likely to be, the best route for the "shouters" who try to drive game toward the guns, and where to station the men with the guns. These meetings also typically involve an exchange of shotgun shells, so that the cost of these items is shared. Following a successful hunt, one of the hunters butchers the animal, and the meat is divided equally among the hunters. This is done using a deliberate process in which the meat is put in even piles equal to the number of hunters, and the piles are numbered with slips of paper. Each hunter searches for the pile that matches the number written on the slip of paper he has pulled out of a hat, and then each puts the meat into plastic sacks carried in his knapsack in anticipation of success. One day when two boars were divided among 17 hunters, I weighed the meat of a hunter who took home seven kilograms of lean, fresh meat, and other hunters presumably got similar quantities according to the egalitarian method for dividing the meat. His household did not eat the pork, however, until negative results came back from a trichinosis test. The hunter who shoots the animal has the right to the trophy from the animal if he desires, although the meat attracted more attention on the hunts that I observed.

Wildlife in Bulgaria is state property, and much of the administration and regulation of hunting is carried out under the auspices of state forestry officials. A hunting council under the national forestry administration makes some of the more important decisions concerning hunting, such as determining hunting seasons, the animals that legally can be hunted, and bounties. During my 1997 fieldwork, the only large game animal that local hunters were legally allowed to hunt was the wild boar, between mid-October and early January. (In most years, they can also hunt roe deer.) Hunting a few game birds was also permitted, but they were not a popular objective for Zaburdo hunters. Hunting of animal pests such as wolves, jackals, foxes, and crows is allowed year-round, and there are bounties for killing wolves and jackals. In addition, most hunting takes place on land managed by the local forest enterprises. The hunters apply to the enterprises for permission to hunt large game in a particular area on a given day. I witnessed a few occasions on which individual hunters declared that they would not go out on the hunt unless the person with the piece of paper giving them permission to hunt was found so that the permission slip could be taken along. In addition, hunters must give some game killed on public forest lands to the forest enterprises. For example, if they kill one boar on a given day, they can keep it; if they kill two, one must be given to the enterprise. If they do not do so and are caught, as happened in Zaburdo a few days before Christmas 1997, they must pay a fine based on the value of the meat from the animal not turned over. In the 1997 incident, a conscious decision had been made to keep both animals, despite an offer from a visiting hunter to transport one of the animals to the forestry office with his jeep, because some hunters wanted the additional meat for holiday meals, and they also expected to go hunting every day of the upcoming holiday week and thus to kill another animal that could be given to the forestry officials. Unfortunately, their decision proved to be costly.

Beyond this, hunting is organized through the Bulgarian Union of Hunting and Fishing and its subdivisions, and all hunters and fishers in the country must be members. The union is the heir to several earlier Bulgarian hunting and fishing organizations that date back to the end of the 19th century. Along with the national-level organization, the union has regional and local structures. A regional society in Plovdiv was established in 1887, for example, and the first regional society in the Rhodope Mountains was established in Chepelare in 1897. The Chepelare regional society includes five local fishing groups and ten for hunting and fishing, one of which is in Zaburdo. The latter is one of two community organizations to which members of surveyed households belong. (The other is the local commercial cooperative.) Membership in the union follows passage of tests of both shooting skill and knowledge of animal biology. Although this organization is independent from the state, some union leaders serve on the state council with authority over hunting. The local hunting and fishing societies may engage in resource management activities, such as maintaining winter feeding stations for birds or animals, improving fish habitats, or suggesting to forestry authorities places where fishing should be prohibited. Some regional and local groups are also involved as nongovernmental organizations in environmental protection activities such as pollution prevention and biodiversity conservation.

Beyond these legal hunting activities, one occasionally hears stories about illegal hunting. For example, a Zaburdo hunter told me that he had found a pair of antlers from a red deer in the forest, and he reported this discovery to forestry officials because hunting this animal was not allowed. It is difficult, however, to determine how much poaching occurs or who is doing it. The regional forestry director in Smolyan said that some poaching takes place in the region but the problem is not large. (He did not provide a specific estimate of how much goes on.) Another question with such illegal activity is the extent to which poachers are local villagers versus the country's newly rich city residents—*Mafiosi* are often implicated in the latter case—with Western four-wheel-drive vehicles and high-powered rifles. Such people are seen as having little respect for the law and may well have faster vehicles and more firepower than the forest police.

Land Restitution and Postsocialist Agriculture

In summing up this discussion of resource use, it is useful to consider the meaning of the postsocialist restitution of agricultural land for local farming and the rural economy more generally. Access to agricultural land is clearly important for the small-scale potato, vegetable, and livestock production that has allowed villagers to survive the last decade. Similarly, the fact that all identified conflicts around natural resources generally concerned the access to, the borders between, and the division of restituted agricultural land demonstrates its importance in postsocialist agricultural activities. It is something worth fighting about. Yet, no one mentioned access to land when I asked people more generally about their problems with or constraints upon such production, and those wanting to expand production could potentially buy or rent additional land. (One man surveyed did complain that a relative had not given him any land, and he consequently did not farm.) Instead, the most frequent production problems identified concerned the affordability and availability of commercial inputs such as artificial fertilizers, modern pesticides, and concentrated animal forage, followed by transportation concerns. Other problems include labor organization, lack of mechanization, low prices for the resulting produce, and marketing difficulties. Put another way, access to agricultural land is a necessary but not a sufficient condition for successful production. One might have plenty of land, but it might not be much use if one lacks the labor, supplies, equipment, or cash necessary for making use of it.

Another point worth making in this regard is that there is virtually no market here for agricultural land. This means that one cannot benefit economically from the restitution by selling one's restored land, although land sales are legal and indeed encouraged by the agricultural ministry. As an economic resource, agricultural land has use value but does not have exchange value. The lack of a market for agricultural land is a widespread problem in much of Bulgaria, but it might be particularly acute in mountainous regions like the Rhodope, where large-scale production is constrained by the landscape and ecological conditions limit the types of crops that can be grown. Small-scale agriculture is an important survival strategy, but the longer-term prospects for more than that are unclear. For example, can smallholder potato production in the Rhodope compete price-wise with large-scale production in Greece or Turkey? Under current production conditions, most smallholder farmers are unlikely to get rich.

Bulgaria's postsocialist restitution of agricultural land has emphasized the "moral" principle of returning land that was taken away to the former owners or their heirs and at the original locations. As such, the restitution promotes an ideology of individuated, private ownership over state or communal property regimes. The material presented here allows us to ask whether the return to private property as an ideal type was the most beneficial form of property ownership or use. Certainly some Bulgarians would disagree. One village friend, a woman whose family has done relatively well in the postsocialist period, has several times commented that she would happily give back the restored land if it meant re-establishing the cooperative farm and with it the possibility of full employment for village residents. In part this comes down to whether people can make effective use of their restored property. Do they have the other resources necessary for accessing the value that the land as a resource has or could have? These are important questions. In the case of agricultural land, it is clear that property access is a necessary but not sufficient condition for successfully making use of it. The lack of real estate markets also plays a role in the question of whether a resource has value. Under current market conditions, land sales are not a viable way to benefit economically from the restitution.

Finally, the farming, gathering, and hunting described in this chapter all take time, and these time obligations must be balanced with other, non-resource-based activities. The next chapter turns to the household economy as a whole and discusses other economic activities and their coordination with those just discussed.

4

Making Ends Meet

This old world of ours is a crazy, topsy-turvy place. There's just no
telling the good from the bad, or how things will turn out in the end.
Nikolai Haitov, *Wild Tales*

This quotation from a short story by Rhodope-born author Nikolai Haitov was written before the political and economic changes
of the 1990s, yet it captures something of the economic uncertainty
of postsocialist life as Bulgarians struggle to make ends meet.
Having a job is good because it provides regular income and presumably a pension eventually; but when inflation turns monthly
wages into the equivalent of pocket change, one begins to question
the logic of going to work every day versus looking for another activity that could generate more income. Forest reprivatization will
presumably benefit people with rights to this resource; however,
the accompanying reorganization of exploitation raises questions
about job security for those currently employed in the forest products industry, as well as questions about the future of the forests.
Private business might be more lucrative than a state-sector job,
but it is not immune from postsocialist economic conditions, and
several village enterprises closed their doors permanently during
the near hyperinflation of early 1997. The 1997 potato crop was a
good one, but cheap imports from Turkey, combined with lower
demand for food following the economic crisis, resulted in lower-
than-anticipated prices for growers. Collecting wild mushrooms

for sale can be quite lucrative, but this business also has a dark side, with secrecy about markets, tales of money laundering, and speculation about the involvement of "the mafia." Both potatoes and mushrooms depend on adequate rainfall, which did not come in 2000, further complicating village economic life.

This chapter focuses on economic reorganization in considering the overall livelihood strategies of village residents in the postsocialist era and the role of natural resources in these strategies. It begins by discussing several aspects of the economic system are somewhat removed from the Rhodope landscape. These include nonagricultural income, barter and other nonmonetary exchange, and the role of cash in the household economy. The second half of the chapter brings together the contributions of different resource-based activities and other income sources to the household economy, painting a picture of how Zaburdo households make ends meet. Following an analysis of the seasonal scheduling of agricultural and other activities, several case studies illustrate a variety of household livelihood strategies, thus highlighting the contribution of resource-based activities to surviving postsocialism in one village. What this means for income inequality is also discussed.

Nonagricultural Income Sources

The dominance of resource-dependent activities in the everyday lives of village residents and their importance compared to other income sources is seen in Table 10, which summarizes the proportion of households participating in various activities that contributed to economic survival in 1996–1997. Compared to households receiving income from potato production, animal husbandry, and the exploitation of natural resources, fewer surveyed households report receiving income that is not directly tied to resource exploitation. (Much of this income is indirectly related to the local landscape, whether it is through employment with the local forest enterprise, the production of handicrafts from locally produced raw materials, or a pension for work on the cooperative farm.) Yet, many households had members who were employed outside agropastoral production, received pensions, earned income from handicraft sales, engaged in business or trade, or some

Table 10. Household livelihood strategies in Zaburdo in 1997.

		Number of households	Percent of sample
Crop production			
Raise crops		118	98.3%
Raise potatoes		117	97.5
Sell potatoes	$	96	80.0
Barter potatoes for goods (answer volunteered)		61	50.8
Raise hay		107	89.2
Own fruit and nut trees, plants, or bushes		56	46.7
Livestock production			
Raise livestock and/or poultry		107	89.2%
Own cow(s)		101	84.2
Own draft and pack animal(s)		67	55.8
Own sheep		30	25.0
Own chicken(s)		26	21.7
Own pig(s)		14	11.7
Own goat(s)		4	3.3
Sell milk	$	100	83.3
Sell wool and/or animal skins	$	25	20.8
Sell live animals	$	8	6.7
Off-farm resource exploitation			
Collect wild products		100	83.3%
Collect medicinal plants, herbs, greens, rosehips	$	88	73.3
Collect wild fruits and nuts		84	70.0
Collect wild mushrooms		77	64.2
Sell wild mushrooms	$	71	59.2
Collect snails (escargot)		19	15.8
Sell snails	$	17	14.2
Heat house with firewood		120	100.0
Household member hunts or fishes		11	9.2
Other income sources			
Pension	$	76	63.3%
Wage employment in last year	$	64	53.3
Handicraft production for sale	$	15	12.5
Other income sources (trade, services)	$	5	4.2

Source: Author's sample survey of 120 village households in 1997.
$ Denotes a potential source of cash income.

combination thereof, and the income thereby generated made a contribution to household economic survival.

Wage Labor

Nonagricultural employment opportunities in the village of Zaburdo are limited, particularly since closure of the socialist-era assembly workshops.[1] In addition, it is rare to see the daily or seasonal movement of workers from village to town, as occurs elsewhere in Bulgaria and other parts of Eastern Europe (e.g., Beck 1976; Creed 1998; Kideckel 1993b). No surveyed households reported members who regularly traveled to jobs in town. Longer-term employment outside the village in the postsocialist period is also infrequent, although a few cases were reported in my household survey. One man had done construction work in the Middle East for a year, another rented out skis at the Pamporovo resort complex one winter, and the grown children of two other households were identified as household members but worked and lived in town most of the year. The latter were said to come to Zaburdo on weekends and vacations to help their parents with agricultural activities.

Beyond these few cases of taking jobs elsewhere, most people relied on the limited employment possibilities with the state forest enterprise, based in the nearby village of Hvoina, or in Zaburdo itself. The latter include a few positions in the service or light-manufacturing sectors at stores and coffee shops, the cooperative bakery, a private dairy, the kindergarten, the school, the local government administration, the post office, the local health service, the sewing workshop (albeit temporarily), and a couple of private sawmills. Additionally, a handful of villagers work at private villas and mountain chalets elsewhere in the village territory or at the tourist hostel on the village square, continuing a long tradition of tourism-related employment in the region. According to my spring 1997 survey data, 53 percent of households had had one or more wage earners at some point during the previous year. This employment did not necessarily cover the entire period, however.

Employment in forestry deserves specific attention, both because of the historical importance of forest products and because of the way in which workers take advantage of their employment conditions to engage in related activities. Bulgarian forests have been exploited by the state since their nationalization in 1947–1948, and the Hvoina-based forest enterprise, whose jurisdiction includes forests around Zaburdo, was the largest single source of nonagricultural employment for village residents in 1997. A frequent sight was forestry vehicles collecting workers in the morning for transport to work sites and then dropping them off in the afternoon. Along with administrative positions, people are employed in both logging and reforestation activities. Logging depends on weather conditions that allow access to forests (e.g., a lack of snow), and tree planting is seasonal, in both cases influencing the regularity with which workers are paid. The gendered division of labor in forestry resembles patterns seen elsewhere. Women more frequently do the hand labor associated with planting trees, while men harvest timber and work with livestock, vehicles, and other machines. By 2001, exploitation activities had been privatized, and a couple of women complained to me about the resulting loss of their jobs as tree planters.[2]

Forestry workers sometimes take advantage of the facts that their work sites are located some distance from Zaburdo and that they have motorized transport to and from these sites; they gather various forest products while at these locations. Among the mushroom collectors interviewed was a group of loggers who had gathered mushrooms on their lunch hour, and on a late spring day that I spent with the tree-planting crew, its members were involved in numerous such activities. One woman collected the new growth from young pine trees to make "pine honey," and several others spent their breaks collecting wood waste to use as firewood. Finally, some workers collected vetch and other weedy plants, which they explained would be fed to their rabbits and chickens. (Another time it was grass to feed their livestock.) Thus, forestry jobs can contribute more than just cash income to household survival. (Taking advantage of the opportunities afforded by the workplace for other activities are not new, as many discussions of the socialist period point out [e.g., Creed 1998; Humphrey 1998; Kideckel 1993b; Ledeneva 1998]).

Although these and other jobs provided households with welcome cash income, on a more regular basis than seasonal potato sales, it is also important to recognize the limitations of these activities. Wage levels are generally low; they did not keep pace with inflation during the 1997 crisis. Also employment is not always year-round or continuous. For example, one young man was employed at a private sawmill only when it had lumber to process, and consequently he took home income on an irregular basis. Tree-planting jobs with the forest enterprise are seasonal (plus the season is the busy summer months), and several people lost their jobs during 1997 when other employers closed or reduced staff. About 15 women were employed by an outside businessman to work in the sewing workshop for a few months in late 1996 and early 1997; however, wage levels were low and the facility quickly closed, putting them once again out of work. At the height of the economic crisis, wages in the jobs that remained were worth perhaps the equivalent of pocket change, yet people continued to go to work on the assumption that things would get better. There were few alternatives in mid-winter, and accumulating enough employment history to qualify for a pension was another concern. Later in the year, some people contemplated taking vacation time to collect mushrooms, however, because it was a more lucrative use of their time.

Handicraft Production

Handicraft production is a source of cash income for about 13 percent of surveyed households—sometimes in the form of hard currency.[3] A few men do woodworking, making wooden crates for seed potatoes or window and door frames for new houses, for example. More commonly, however, women weave tufted blankets and rugs using locally produced wool and goat hair, or they knit or embroider slippers. These items are commonly used in village homes and are a significant component of the dowries that women take with them into marriage. In some cases, textiles or slippers are produced directly for sale; in other cases, women sell pieces from their often large dowries when there are more items than could be needed for household use (see also Creed 1998:203–204 on Bulgarian dowries). Production for sale was organized through an artisan cooperative during the socialist era, but that system is no longer in place. Yet, Zaburdo women are well known for this type of production. Potential customers sometimes come to the village in search of these items, and a few larger producers transport their goods to Pamporovo, Plovdiv, or Sofia for sale.

The Village Business Enterprise

Members of a few households earn income through private business, trading, or providing services. Only about 4 percent of surveyed households reported such activities, although the proportion for the village as a whole may be somewhat higher given the number of coffee bars, mushroom buyers, and villagers involved in trading goods for potatoes. For some people, these activities are in addition to regular jobs, such as a carpenter for a private firm who does simple electronic repairs on the side. In other cases, an individual's or a household's main nonagricultural activity is running a store or a coffee bar, buying mushrooms, or transporting and trading goods. Sometimes, the same person or entrepreneurial household engages in several different business activities as the opportunities present themselves, either serially or concurrently. One entrepreneur, for example, runs a coffee bar year-round and seasonally buys mushrooms and rosehips. In previous years, he had been involved in other activities such as construction. Such business ventures may generate more income than a state-sector job, yet most people also participate in or at least belong to a household engaged in farming. The share of the population involved in business is limited, and these activities have not been immune from the inflation and other problems that have plagued Bulgaria's economy for several years.

Take, for example, the case of Georgi and Eli Petrovi. They opened a store selling food and household goods after the dust settled from the 1997 crisis, and for a time it was the brightest and best-supplied store in the village. By spring 2000, business had declined, and a newer store seemed to be attracting more customers. Upon seeing me in the village after a two-year absence, Eli commented on the fact that Bulgarians are all desperate to get out of the country due to the economic situation, while I seemed to keep coming back. "Are you crazy?" she asked me. Meanwhile Georgi had built a small sawmill and hoped to get into the wood-processing business. He asked me to be on the lookout for equipment suppliers in Germany, where I lived at the time, thinking that equipment there would be cheaper than it was in Bulgaria. By 2001, Eli had given up the store in the village and was working in town,[4] and Georgi had sold his wood-processing machinery, explaining that he could not make a profit after paying taxes and various employment-security contributions. Last I saw him, he was buying wild-collected mushrooms in the space that had once housed the store, something he had done in other years as well.

Pensions and Other Financial Assistance

Finally, some residents also have income from pensions and financial or social assistance. According to my survey data, 63 percent of households receive income from disability pensions or (more frequently) old-age pensions, and 38 percent receive two or more pensions. This is a higher percentage than those households receiving wage income, and this income source may provide a more continuous stream of money into household coffers than temporary or seasonal jobs. Monthly pensions are not large—in the range of 4,000 to 10,000 leva ($1.30 to $3.30) during the early 1997 economic crisis and 30,000 to 80,000 leva ($17 to $45) later in the year after the economy had stabilized. Some people would say, for example, that with a pension they could barely afford a month's supply of bread. Yet, the money received is important for paying for things that cannot be obtained through barter with potatoes, and it is a critical resource for people too old to work and without someone to assist them with agricultural production.[5] Thus, some say that with one or two pensions, they can buy bread and pay their utility bills and that other than the purchased bread they can eat what they produce themselves or can obtain through barter.

Indeed, pensions are a more important consideration than I expected, and this is true for working-aged people as well as retirees. During the height of the 1997 economic crisis, for example, a 40-year-old forestry employee explained that his monthly salary was worth only about "three dollars" (he did the conversion to dollars himself, as was common practice at the time), but that he kept his state-sector job because it meant that he would eventually receive a pension. (His salary had regained some of its value by the end of the year and was worth about $70.) Other people I spoke with also cited the eventual receipt of pensions as their reason for keeping low-paying jobs. Similarly, a frequent topic of discussion among people nearing retirement age—whether or not they were employed—was their number of years of service (*trudov stazh*) and how many more years were necessary for pension eligibility. This is a particularly vexing problem for those people nearing retirement age but unemployed and with little prospect of employment in sight given the postsocialist job market, especially in rural areas.

Beyond pensions, some households or members thereof receive unemployment benefits, one-time financial or material aid, or a combination thereof. Due to the one-time nature and timing of this assistance, my survey data provide an incomplete picture, and it is more useful to talk about the kinds of assistance and the circumstances surrounding its receipt. One occasionally hears about unemployment or other benefits from the state sector. One woman who lost her state-sector job during my fieldwork opted to receive her unemployment compensation in a lump sum because the household needed money at the time. Through a government program for unemployed residents, several people were employed temporarily by the local administration as street sweepers or general laborers. While wages were low, these several months of work did count toward pension eligibility. The money was undoubtedly a welcome addition to the household budgets as well.

Along with these government benefits, one-time or occasional financial and material assistance from non-Bulgarian sources made its way to village residents in the aftermath of the early 1997 economic crisis. A shipment of flour from Greece resulted in reduced-price bread at the bakery for a time, and oranges in the same shipment were distributed to households identified as being particularly needy. On a couple of occasions, needy villagers applied for cash assistance payments from the European Union, which some then received. On other occasions poorer or otherwise eligible villagers obtained through such programs, reduced-price, concentrated forage for their animals or free coal for heating their homes.

These forms of aid were not without controversy or conditions, however. Some of the oranges reportedly spoiled while waiting for shipment at the Greek-Bulgarian border, and there were rumors that others were sold for the benefit of people who were not in need. Similarly, the unavailability of applications for and subsequent "misallocation" of the European Union funds to people "with money," while "poor grandmothers went without," was a common topic of discussion among people ranging from the grandmothers themselves to the village doctor and mayor. On other occasions, local residents questioned the way the aid was delivered. The recipient was responsible for paying for a portion of the transportation costs for the free coal, for example, and an 80-year-old grandmother told me that she could not afford to do so given her small pension. In another memorable case, an acquaintance explained that his 80-year-old mother had received a coupon for a 100 leva discount ($0.06) on each of three loaves of bread (the regular price was 600 leva per loaf, about $0.33). In his opinion this was an absurdly small benefit. Why couldn't they at least have given her a coupon for one free loaf of bread, he asked. In the end, he tore up the coupon in what he described as a protest. These examples illustrate the somewhat episodic and unexpected ways in which financial and material support arrived in Zaburdo, perhaps assisting the recipients somewhat with a difficult financial situation but not necessarily in a sustainable way or in the most helpful fashion.

Features on the Postsocialist Economic Landscape

Although nonagricultural income sources are clearly important, a more complete understanding of the postsocialist economic system in the village is gained in discussing three other topics: barter, other nonmonetary transactions whose purpose is more than the simple exchange of goods, and cash-based transactions. In some sense, they represent three points on a continuum between gift giving at one end and impersonal commodity exchange using cash at the other, and they also illustrate how the formal and informal sectors continue to exist side by side.

The Basics of Barter: "You Can Buy Almost Anything with Potatoes"

Barter is a significant and visible feature of the postsocialist economic landscape in Zaburdo and other Rhodope villages. The exchange of produce or firewood for land and labor as part of the agricultural production system is just one example of the nonmonetary exchanges of goods and services that take place in such villages. On one of my exploratory visits to these mountains while looking for a field site, one village mayor explained that one could acquire almost anything in exchange for potatoes—grapes, oil, sugar, concentrated animal feed, fertilizer, the list went on— and he expressed a desire that his office help organize the exchange in a more efficient fashion for the benefit of local residents. As time went on, it became apparent that such barter was common, often taking place alongside cash sales, and that several different kinds of transactions were occurring. Before discussing exchanges of goods without the direct use of money in Zaburdo, a few words are in order about barter more generally.

Barter and other forms of exchange have been important topics for anthropologists like Malinowski (1922) and Bohannan (1955), as well as more recent scholars like Humphrey and Hugh-Jones (1992); also Humphrey (1985). Combining several recent definitions, Appadurai (1986:9) defines barter as "the exchange of objects for one another without reference to money and with maximum feasible reduction of social, cultural, political, or personal transaction costs," noting that the first criterion distinguishes barter from commodity exchange, and the second distinguishes it from gift exchange. Elsewhere in the essay, he notes the possibly indirect role of money as a unit of account in evaluating barter transactions, as well as the fact that barter cannot be entirely divorced from sociality. Other recent anthropological discussions of barter raise similar issues concerning the difficulty of defining barter in opposition to other forms of exchange and in a way that covers all forms of barter (e.g., Barnes and Barnes 1989; Humphrey and Hugh-Jones 1992).

In the introduction to a 1992 book on the subject, Humphrey and Hugh-Jones (1992:1) suggest that common features often shared by exchanges known as barter include: (a) demand for goods or services that are different in kind, (b) free and equal participants who can pull out of the deal at any time and have no future obligations following completion of the exchange, (c) an absence of outside criteria by which the items being exchanged can be judged as being equal in value, or the use of abstract value in the negotiation process—each of the parties involved simply wants the object held by the other and is willing to give up something he or she has in order to obtain it, (d) the two parts of the transaction may occur simultaneously or separated in time, and (e) the act of the transaction moves the objects between value systems. Having laid out these characteristics for an activity that is described as lying somewhere between gift giving and commodity exchange, the authors caution against using this as a checklist or universal definition. Instead, they emphasize the importance of the social and cultural contexts of barter, on the one hand, and way in which barter relates to other forms of exchange, on the other.

Several other points raised by Humphrey and Hugh-Jones (1992) are useful to keep in mind in considering the Zaburdo case. Participants in barter exchanges may have different evaluations of what is going on in a single transaction. For one participant, the transaction might be an exchange of goods for goods, neither of which is measured in terms of money; while for another participant in the same transaction the barter might be seen as a crude substitution for money. The authors also point out the coexistence of barter and the modern world economic system, with examples of the exchanges between socialist and capitalist countries before 1989, many of which took place entirely in goods, as well as the barter networks that are popular in capitalist economies today (Roha 1996 provides an example). Indeed, Humphrey and Hugh-Jones (1992) write that barter may occur in several different circumstances: in the absence of money, alongside a common currency that people prefer not to use for whatever reason (e.g., to avoid taxes), and when there simply is not enough money to go around. Thus, barter may serve as a solution to the problems of money, rather than the other way around—that is, the more conventional view that money is a solution to the problems of barter.

Humphrey and Hugh-Jones (1992) also stress the importance of the social relations involved in such exchanges (see also Barnes and Barnes 1989; but cf. Chapman 1980, who sees no role for social ties in pure barter). In particular, relationships in simultaneous barter are discontinuous and unstable in that no additional transaction is needed to satisfy the wants of the actors following completion of the exchange. In other words, once the transaction is completed with the exchange of objects, it is not necessary for the parties involved to have further contact or interaction. For such transactions to occur, however, information is needed about what can be traded, where, when, and by whom. Without a "double coincidence of wants" (Pryor 1977:158, cited in Orlove 1986:85)—that is, two actors, each wanting what another has and

being willing to trade—barter is unlikely to occur. Such information is often a critical factor in locating or arranging a coincidence
of wants. The possibility for repetition leads to a need for fair
dealing in most instances in order to establish a reputation for
trustworthiness and assure the partner that such a transaction
could be repeated in the future. Yet, fair dealing is not always
guaranteed. Merchants may employ in-kind payments to keep
people out of the money-based market, or they may hide information about the outside prices of the goods being offered in
order to have greater control over the prices that they charge.
Humphrey and Hugh-Jones (1992:11) suggest that such exploitation is maintained by conditions external to the exchange, such as
existing inequalities or a lack of market information. Yet, the possibility for exploitation suggests the need for a careful examination of the motivations of the participants involved in barter exchanges and the circumstances surrounding the transactions.

In the last decade, barter has captured the attention of scholars
of the postsocialist world, although the small-scale transactions
between producers of the kind that I observed have received less
press than the barter that takes place among large enterprises,
and much of the focus has been on the former Soviet Union rather
than Eastern Europe (e.g., Woodruff 1999 and the contributions
to Seabright 2000). An account by anthropologist Caroline Humphrey (1998:chap. 9) of a post-Soviet collective farm in rural Russia describes how workers are paid in goods that they do not produce by bankrupt enterprises, and then the somewhat surreal
way in which such goods (including, in one instance, ten thousand saplings) circulate through the economy until they end up
in the hands of someone who can use them (Humphrey 1998:471).
These elements are repeated in many journalistic accounts of the
Russian economy, which often focus on the most bizarre items
that have been used to pay employees (e.g., *The Economist* 1997)
and the prevalence of business by barter more generally (e.g.,
Caryl 1998; Goldman 1998). One reason for the lack of attention to
smaller-scale barter may be that it was not unknown under the
"economy of shortage" (Verdery 1991), which characterized the
region during earlier periods, and thus was nothing new to get
excited about. Under socialist-era conditions, in which prices
were not necessarily governed by supply and demand and
both consumer goods and industrial supplies were scarce, the ex-

change of goods without the use of money, by both individuals and businesses, was not uncommon.[6]

My conversations with Zaburdo residents suggest that barter played a relatively small role in that village's economy during the socialist era. Most villagers' descriptions of this period focus on the cash nature of the economy—the money they earned, what they spent it on in the village store, the price of bread, and so on. When I specifically asked some of my village acquaintances about barter before 1989, one common response was, "Why would we have done so?" They had money, and there were things to buy in the well-supplied village shop; and if they wanted to go shopping in town, public transportation was frequent and inexpensive. Upon further investigation, a few people mentioned exchanging potatoes for forage with the cooperative farm, while others explained that occasionally someone would come from another village to barter small quantities of something that they had produced themselves.

This situation had changed by the time of my postsocialist-era fieldwork. My first experience with the most widespread form of barter I observed in Zaburdo—the exchange of potatoes for goods—occurred during the midst of the economic crisis in early 1997, when I went with a friend to exchange (*smenyam* in Bulgarian, meaning to change or exchange, although the English word "barter" is occasionally used) small seed potatoes for flour from a man who had arrived in the village with a truck full of bags of this staple good. But the timing of the experience may have been more a factor of when I arrived in the village than the specific economic conditions of the day. The practice occurred both before and after this time of particularly serious economic instability and often alongside the sale of potatoes for cash.[7] In some cases, a "seller" would announce prices for a good both in potatoes and in cash; in other cases, two separate vendors, located side by side, might provide these two alternatives. People seeking to acquire potatoes in exchange for goods usually arrived with an exchange rate or price in mind, for example, 1 kilogram of apples for 1 kilogram of potatoes, 3 kilograms of seed potatoes for 1 kilogram of carrots, or 20 kilograms of potatoes for several packages of rice. Although somewhat negotiable, as an example below illustrates, this price often appears informed by the money values of potatoes and the goods for which they are being exchanged. In discussing his socialist-era fieldwork, Creed (1998:206) similarly observes that barter involves assigning a monetary value to the items exchanged and thus is not necessarily evidence of decommodification of the economy or the disappearance of a cash nexus (i.e., counter to point [c] of Humphrey and Hugh-Jones [1992] as discussed above, but see Humphrey 2000). Indeed, in the Zaburdo case at least, the potatoes sometimes seem to serve as a proxy currency—albeit a lumpy one—in the absence of cash and thus function as a solution to some of the problems of Bulgaria's postsocialist market economy, without necessarily being encumbered with the multistranded connections sometimes described in relation to barter. This is further illustrated in the examples below.

Following the potato harvest and through the winter months, a frequent piece of information passed along the neighborhood grapevine relates to the items available in the square in exchange for potatoes, and people keep watch out of their windows to observe the contents of cars and trucks entering the village. One day a group of women from my neighborhood besieged a man from a nearby but lower-elevation village, who arrived with a car full of walnuts. After telling him that the person he had brought the walnuts for could not possibly want all that he had, they bargained about the price and then quickly collected the necessary potatoes to buy two large gunnysacks full of walnuts. Another day several men were standing around the square eyeing a truck full of concentrated animal feed. An acquaintance in the group explained that the asking price was too high and they were holding out to get more grain for their potatoes. A short time later the men scattered to bring their potatoes to the square, so a deal must have been struck. Although it had not occurred to me to ask, about half the surveyed households volunteered that they exchanged potatoes for goods when asked about their potato sales, and thus the actual proportion of households doing so is likely even higher.

A household's level of involvement in barter at a given time depends upon at least two variables: the opinion about such a deal and the availability of cash with which to make purchases.[8] A retired widow said that she was most comfortable trading potatoes for goods. Conversely, the adult son in a successful household said that his household rarely barters because a better economic deal is obtained by selling potatoes for cash and then using the cash to buy other goods. This household is in perhaps a better position to do so because it has regular infusions of cash from nonagricultural activities, and it also owns a truck that can be used to transport potatoes to market. Such opinions appear to reflect an evaluation of the economic rationality of transactions and perhaps also the morality of such exchanges. Most people appear to fall somewhere in-between—sometimes bartering, other times making purchases with cash. In discussing barter between pastoralists and agriculturalists in Greece, Forbes (1997) points out that in exchanges of products respectively produced and consumed by the parties involved in the transaction, an economic advantage can accrue to both parties when the producer can obtain a price above the typical wholesale value while the consumer pays less than the retail cost.

In Forbes's (1997) example of the exchange of cheese for the use of pasture, another issue raised is the value placed on the quality of a specialty or artisanal product obtained from a particular producer. This may also have significance in Bulgaria, where there is a lack of trust in the quality of items purchased on the open market. Anxiety about the health hazards from counterfeit alcoholic beverages is particularly evident, but such concerns extend to other products as well. People question, for example, whether commercially available milk or butter has been diluted with water or oil, the source of ground meat, and whether products are fresh. Consequently, one factor potentially influencing decisions to engage in face-to-face barter for a product with its producer, rather than buying it for cash in an anonymous store, is a greater degree of trust about the quality of the product thus obtained (see also Barnes and Barnes 1989; Kipnis 1997:153; and Smart 1993 on the role of trust and social ties in facilitating economic transactions in uncertain regulatory contexts). Clearly, trust would be facilitated by prior ties between the parties involved, but the face-to-face nature of transactions or the fact that transactions take place between residents of neighboring villages with a history of trading might play a role in the absence of specific ties between individuals.

While various opinions about this form of transaction play a role in explaining the level of barter I observed, access to cash combined with the underdevelopment of postsocialist marketing systems is perhaps more important. For households without cash, barter is a way to obtain goods and thereby to deal with one problem sometimes encountered in the postsocialist market economy. With the liquidation of the cooperative farms and breakdown of the state-controlled food distribution systems, produce marketing has become a problem for rural farmers. As discussed in chapter 3, most residents sell their potatoes in the village, in some cases due to lack of a vehicle with which to transport the bulky produce to market, and they are consequently dependent on a buyer to come to them. During fall 1997, the potato market was saturated due to a good harvest and reduced consumer demand resulting from the economic crisis. (The availability of cheap imports from Turkey was also cited as a factor by some.) Traders did not initially come to buy potatoes for money, although a variety of goods were available for exchange. Since villagers had no money but many potatoes and a desire to make purchases, a considerable amount of barter occurred.

The following story, related to me by a young woman, illustrates this point and also shows the substitutability of potatoes for money. One fall morning a few days before our conversation, the young woman had noticed that a man, who she described as a Bulgarian Turk from the eastern Rhodope town of Kurdjali, had arrived with carpets and clothing to sell and had set up shop in the village square. She told him that people would not buy his goods with money because they did not have any, since they had not yet sold their potato crop, and she suggested that if he sold the goods for potatoes instead, he would sell them all. He tried to sell them for money for a little longer, she said, but eventually, after some negotiation over the money value of the potatoes versus the goods, he offered for sale a women's blouse for 50 kilograms of potatoes, a leather jacket for 130 kilograms, and so forth. He managed to sell most of his goods by the end of the day, she reported. The parties involved in this example seemed to be simply responding to the circumstances in which they found themselves, substituting potatoes for money.

I sometimes wondered, however, whether some traders were taking advantage of the lack of cash in the village and trading goods at higher-than-market prices to a captive audience.[9] This is suggested, for example, by the fact that traders came to the village to barter goods for potatoes in large quantities after the 1997 harvest, but few initially came to buy potatoes for cash. Villagers' observations regarding the prices also hinted that this might sometimes be the case. Some village residents commented that generally the prices available in the barter transactions are not favorable, and others made similar remarks about particular prices being too high—such as one friend's comment that three kilograms of potatoes was too much to pay for one kilogram of carrots.

While it seemed problematic to investigate directly the extent to which traders might be manipulating the producers, the possibility prompted me to explore the larger social and economic contexts of such transactions in a kind of "thought experiment." In particular, it is useful to consider the circumstances of the transactions in terms of the goods exchanged, the identities of the traders involved, the connections and information sources going into the system of barter, and the power relationships that such transactions entailed. During my fieldwork, Zaburdo residents reported obtaining the following goods through barter: cabbages, carrots, leeks, onions, peppers, tomatoes, apples, bananas, grapes, pears, oranges, tangerines, dry beans, flour, lentils, oil, rice, salt, sugar, walnuts, dishwashing soap, laundry detergent, tablecloths, concentrated forage, fertilizer, socks, shoes, jackets, sweaters, stirrup pants, grape brandy, and cigarettes. While this list certainly supports the earlier-cited mayor's claim that you can buy almost anything with potatoes, the other thing to note is that not all of these products could have been produced by the person trading them or by a friend or relative. Several items are clearly commercial products and in some cases from other countries.

So, one might ask, how is this trading organized, and who is doing it? With the breakdown of the socialist-era systems of produce purchasing and distribution, there remains general knowledge about the kinds of products produced in different parts of the country, such as potatoes in the mountains and grains and vegetables in the plains, and thus the possibility of purchasing or otherwise obtaining such goods by going to particular places. Finding a "double coincidence of wants" is not necessarily an insurmountable problem when it comes to agricultural products. In some transactions, the trader has come on such a trip to Zaburdo before; in others, the potential potato buyer is a complete stranger. Similarly, while villagers might not know specifically who will come to buy their potatoes, they operate on the assumption that there is a demand for this product and that someone will come. Indeed, because of the lack of transportation resources to take the

products to a larger market, many villagers are a captive audience
for such itinerant traders. Thus, contrary to the case described by
Humphrey (1992), in which the host largely controlled the price,
these villagers may have only limited bargaining power.

The list of items for which one can trade potatoes suggests that
there are at least two types of buyers and two types of goods in
such exchanges. In some cases, the people coming to the village
with modest amounts of goods to trade for potatoes have produced the goods themselves. Examples of this include the walnut
"seller" described earlier, along with another man who had come
from Asenovgrad to barter feed corn that he had grown on land
belonging to his wife. This conversation is still clear in my mind,
as he said that this land was not the reason he had married her—
private property was irrelevant at the time of their socialist-era
wedding—but that it was clearly to his benefit now. These transactions are reminiscent of socialist-era exchanges between households of goods not necessarily available through the state distribution system. They also could be examples of situations in
which—as Forbes (1997) suggests—an economic advantage ac-
crues when producers and consumers exchange goods without
the involvement of an intermediary trader, or when these transactions are facilitated by a greater degree of trust regarding the
quality of the products exchanged. In other cases, however, the
item available for barter is a manufactured one, such as fertilizer
or sugar, or one that does not grow in Bulgaria, such as bananas.
In such instances, several different goods were often available
from a single trader. Some villagers serve as such traders, acquiring goods somewhere, transporting them to Zaburdo, and then
exchanging them for potatoes, which they transport elsewhere.[10]
Another difference in the nature of the exchanges often coincides
with the different origins of the bartered goods. Most producers
at my field site, as well as the other small-scale producer–traders
of walnuts, apples, feed corn, and so on described above, barter
for relatively small quantities of goods, either for their own consumption or for sharing within a circle of family and friends. In
contrast, larger-scale traders sometimes obtain large volumes of
potatoes (e.g., several metric tons), which are probably not destined solely for their own households' use.

It is possible that the larger-scale traders who barter commercial or nonlocal products for potatoes are in a position to obtain

such goods through informal connections (whatever their place of residence). But this situation also suggests the need to consider the motivations and strategies involved in such barter and the way in which it is both facilitated and constrained by economic conditions in Bulgaria. Of particular interest might be the specific sources of the goods brought to the village and the disposition of the potatoes acquired for these goods. Are the potatoes, for example, next exchanged with an end consumer, an institutional customer, or an intermediary trader? What is the relationship between the traders who acquire the potatoes in the village and people or firms acquiring the potatoes from the traders? And do these subsequent exchanges involve goods, money, or both? Also, is the primary goal of traders to obtain potatoes or to sell other goods? And, related to this question, would traders sell the goods for money if the producers had the cash to make purchases? In other words, are they engaging in barter because the villagers lack cash? Or alternatively, are they bartering, regardless of the cash supply in the community, in order to avoid the use of money for other reasons? I am not in a position to answer these questions, in part because my research focused on local production issues rather than the distribution system per se. Yet, it seems that under postsocialist conditions, barter is facilitated by a cash shortage and problems with the system for buying and selling produce, and that limited, effective controls on petty trading may create opportunities for entrepreneurs to exploit the situation.

Finally, how did the hyperinflation of early 1997 affect the preexisting system of barter? Undoubtedly the volume of nonmonetary exchange increased as inflation skyrocketed and formal institutions of economic exchange such as retail stores ceased to operate. But one still might ask whether the barter was related to the inflation itself or to other factors, such as an absolute shortage of goods caused by hyperinflation and economic chaos. While the local currency was spiraling out of control, at least two generally recognized and relatively stable external currencies—the U.S. dollar and German mark—were available at exchange offices around the country and were used in some settings. For example, gasoline, in short supply because Bulgaria had been unable to pay its bill to the supplier of its one refinery, was reportedly available on the black market in exchange for dollars or marks. Even if one had money, however, it could not always be spent because goods might not be available for money,[11] and thus it was necessary to resort to barter. An example of this comes from a friend who worked for a transportation company in a small Rhodope town. She and her boss, both smokers, arranged to exchange cigarettes for gasoline from the company's tanks, which they later refilled at the new price. The important thing to note about this transaction is that both of these goods were unavailable in formal markets. And the larger point is that barter activities in postsocialist Bulgaria are complex and multifarious, involving issues of the organization of systems of distribution, access to cash, values placed by specific individuals on different forms of exchange, and information availability.

Other Forms of Informal Exchange

Besides the direct exchange of agricultural products such as potatoes and milk at specified rates for goods or land rent, Zaburdo residents also engage in less formal and less direct—albeit no less institutionalized or important—forms of exchange. As Creed (1998:205) points out, these exchanges generate obligations and long-term connections that have value beyond the products exchanged. Humphrey (1992) and Humphrey and Hugh-Jones (1992) also discuss the social context and social relations involved in nonmonetary exchange, and the relations that they describe are more apparent in this type of exchange in Zaburdo than in the previously discussed barter with potatoes. When friends and children living in town visit Zaburdo, village households often send them home with potatoes, milk, other dairy products, home-canned fruits and vegetables, and firewood. The villagers also bring these items as gifts when they visit family or friends in town or the lowland villages. In turn, the recipients in town may send to the villagers (or help them acquire) items not readily available in the mountains, such as tomatoes, cabbage, grapes, concentrated forage, and even grape brandy. Labor may be exchanged as well, when the friends and children from town visit to assist with agricultural activities.

But these exchanges extend beyond the family. The high-quality dairy products and livestock from Zaburdo are an important and universal currency for obtaining special favors or services or for sealing particular deals (see also Ledeneva [1998] on "the economy of favors" in Russia). Two villagers, for example, each related giving sheep to doctors who had taken care of them in particular moments of need. I also accompanied a friend one day as she searched for locally produced cheese to take to a doctor she needed to visit in town. On another occasion, one of my Zaburdo *blizki* asked me to take his young daughter to study medicine in the United States when the time came, explaining how difficult it was to get into medical school in Bulgaria and how well she did in school. In return, he said, he would regularly send me milk and cheese from his cows, to which—as he knows—I am particularly partial.

These practices seem rather matter of course to me after spending more than two years in Bulgaria, and Ledeneva describes similar practices in the Russian context as "nothing special at all—just a daily routine, habitual" (1998:4). Grodeland et al. (1998) discuss such exchanges in return for medical care or other services in terms of bribery and corruption, however, and note that their research uncovered numerous examples of participation in such exchanges by the Bulgarians who they interviewed. Some examples they discuss clearly involved a direct or indirect request for payment, and I was also aware of such cases. An example of this was seen in chapter 1, in the difficulties experienced by one NGO in obtaining a telephone line for its office. But in other instances, exchanges of meat or cheese, for example, are in a sense voluntary, albeit institutionalized and important for establishing and maintaining networks of connections—connections that continue to be important. Another point about these practices more generally, given what may be a misunderstanding of them by those unfamiliar with their embeddedness in society, is the possibility that they will influence other sorts of social relations in the postsocialist period, including those affecting NGOs (e.g., Bruno 1998; Ledeneva 1998; Wedel 1998a:9; Werner 2000).

The Importance of Cash

Concerns about earning a pension, as well as poorer residents putting off their own agricultural work to earn day wages by working for others, reflect the importance of money in the postsocialist village economy. While many items can be obtained through barter, villagers must have cash to pay telephone and electricity bills and to buy bus tickets, bread from the bakery, and modern appliances such as televisions and VCRs. Having money also affords one the freedom to shop around in village stores or in town at one's leisure, rather than waiting for vendors interested in exchanging the desired goods for potatoes and then making do with the limited selection available. In discussing post-Soviet rural Russia, Humphrey (1998) points out that the importance of cash lies in its instant convertibility, even during times of rapid inflation. As she notes concerning pensions in Russia, and as I also observed in Bulgaria, cash might be immediately spent upon receipt to avoid losing too much value to inflation, but having money has particular value.

Large-scale participation in the cash economy was a socialist-era development in Zaburdo. During that period most villagers had jobs that paid in cash—in contrast to the more subsistence-oriented economy of private times—and there were also places where villagers could spend their money. In discussing the socialist period, Zaburdo residents rarely referred to the time of shortage in Bulgaria, during which there was nothing to buy even though they had money. (This latter situation was a frequent refrain when I spoke with people in other communities: "It used to be that we had money," people would say, "but there was nothing to buy." "Now you can buy anything," they would continue, "but we have no money.") Instead, Zaburdo residents reminisced about how, during the socialist period, local stores were full of goods, brought by truck three times a day, and about how low the prices were.

Utility bills, bus tickets, and VCRs may not have been "necessities" during private times, and bread was baked at home before the bakery opened in the early 1960s. Many households became accustomed to paying for these expenses with the cash earned during the socialist era, however. Indeed, the introduction of a widespread cash economy in the village and the availability of goods to purchase with cash may have led to the discontinuation of some household production processes, as a couple of examples illustrate. Although some people continue to prepare salted and dried meat (*pasturma*) and a dried soup-base they call *liuta kasha* (lit., spicy porridge) at home, one woman in her mid-fifties told me with some regret that she missed the molasses and dried fruit that were produced in her childhood. When I asked why these items were no longer prepared, she said that most people had stopped doing so during the socialist period because they had money and there were things they could buy (e.g., sugar instead of molasses). Although she continues to grow the sugar beets that earlier had been used to make molasses, she does so for the leaves, which are eaten like spinach, and feeds the tubers to her cows. Equipment for processing the sugar beets into molasses has disappeared, she said. Now even the poorest households regularly line up at the bakery to buy bread, which is an important staple food in Bulgaria, and at the post office or other payment location to pay their telephone and electric bills.[12] As well, buying modern colored televisions, chainsaws, and livestock requires large cash outlays.

Cash is also important for the education of children. Sending children to secondary school became a widespread practice after the creation of the cooperative farm in Zaburdo and increased incorporation of the community into the larger political and social system. Tuition is not charged for public schools; however, sending a child to high school in town (since the village school only goes to the eighth grade) requires a family to pay for housing, food, books, and periodic transportation for the children to return home or for the parents to go to town to visit the children. These visits are for more than pleasure, as the luggage of the child or parent traveling between village and town is usually filled with jars of home-canned food to sustain the student until the next visit (see also Kaneff 1998:23). Several parents complained to me that sending children to school was a significant drain on their resources, even with both parents working. In a couple of cases, village young people of my acquaintance wanted to continue their educations beyond high school but due to the costs involved had been unable to do so, at least not directly following graduation. Education is identified as being important for access to nonagricultural employment opportunities, and Humphrey (1998:462) similarly cites concern by families in postsocialist rural Russia about financial difficulties being a barrier to higher education for children. Both cases also reinforce the point that such opportunities were available during the socialist period, even to people in rural settings. Even if some villagers are (once again) primarily subsistence farmers and obtain most items that they do not produce themselves through barter or other forms of nonmonetary exchange, some degree of articulation with the cash economy is still necessary.

Meanwhile, some villagers seek to decrease the use of things for which they must pay money in response to the cash shortage and increased prices. Increased participation in subsistence food production and reliance on firewood for heating have been discussed, but there are other examples as well, and they show the ends to which people will go in this regard. The ways in which people have sought to reduce their electricity use illustrate their desire to decrease their dependence on money. Many households own electric devices for churning butter, but one day a forestry worker regaled her coworkers and the anthropologist who had tagged along for the day about a hand-powered churn she had recently obtained in a nearby settlement. She said that you could sit and watch television while cranking the churn, thus avoiding the use of electricity in butter production. Another person unplugged his VCR when it was not in use to avoid using the electricity that powered the clock and a small light inside the unit. He continued to use the appliance, plugging it in when he used it, but sought to reduce unnecessary energy expenditures while it was not in use.

Overall Household Livelihood Strategies

The discussion thus far has sketched out the individual components of household provisioning and economic activities. Village residents engage in multiple activities dependent on the surrounding landscape including agriculture, pastoralism, gathering, and hunting. They also receive income from sources that are not directly based on natural resources, and traditions of barter and informal exchange exist alongside the use of cash. The remainder of this chapter pulls together these individual components into a more complete description of how village residents are making ends meet and perhaps even trying to get ahead in the postsocialist period. The discussion begins with the seasonal organization of labor, moves on to several case studies of actual households, and then describes perceptions of economic stratification.

The Seasonal Scheduling of Activities

An important consideration for the household economy is the annual timetable of labor and activities. Figure 5 summarizes the seasonal scheduling of the resource-related activities in which Zaburdo residents engage. This seasonality is influenced by environmental conditions in this mountain setting. Some activities take place throughout the year or can occur whenever the weather cooperates and time is available. Basic animal care, for example, takes place year-round, and goats and sheep are taken out to graze by some households in the winter when snow conditions permit. Hunting is almost exclusively a winter pastime and thus does not particularly conflict with other resource-based activities. Similarly, some field preparation and transport of manure can take place in the late fall and early spring, and firewood is often collected during lulls in potato production and then between the potato and hay harvests and the onset of winter snows. Thus some activities do not present particular problems for scheduling of household labor for resource-related activities, although time still must be found for them.

Many activities are concentrated in the summer months, however, and conflicts over household labor for agriculture and other resource-related activities are most likely to occur during this season. (Cole and Wolf [1974:128] discuss a similar annual pattern of activity in the Italian Alps.) Although wild fruits may be available for the taking, households might not be able to pick them if they are hurrying to get their hay harvested. One day I was left at home to finish canning cherries over a wood fire in the front yard, while my landlady rushed off to bring some hay in before it rained.[13] Similarly, young adults might be criticized by their parents for hunting mushrooms every day. They are drawn by the prospect of immediate cash payment, but this also means that they neglect other activities seen by elder household members as critical for getting the household through the winter, such as cutting enough hay to feed the animals through the winter, thereby securing

	JANUARY	FEBRUARY	MARCH	APRIL	MAY	JUNE	JULY	AUGUST	SEPTEMBER	OCTOBER	NOVEMBER	DECEMBER
POTATOES	*(weather permitting, field preparation and transport of manure to field, also sorting of potatoes if not completed)*			plow, plant, and fertilize; apply herbicides 2–3 weeks after planting		1st cultivation, fertilize	2nd cultivation |- pesticides -|		harvest	sort potatoes *(weather permitting, field preparation and manure transport)*		
VEGETABLE GARDENS	*(weather permitting, field preparation and transport of manure to fields)*				|- plot preparation and plant -|	garden care and harvest occurs as needed and as time is available						
HAY PRODUCTION	*(weather permitting, transport of hay from storage facilities near hay meadows to barns where animals are cared for)*				|--- fertilize (rarely) ---|	|- first and sometimes second cutting of hay *(may cut green in early June if need food for livestock)* -|			*(weather permitting, transport from storage near meadows to barns with animals)*		
COWS							|- communal herding -|			|- individual grazing -| *(weather permitting)*	
	(Calves may be born at any time of the year. Cows are milked year-round when lactating.)											
SHEEP	|----- year-round grazing in pastures around village, weather permitting (i.e., barring snow), animals move into potato and hay fields after harvest ----|	|--- lambs born ---|		|-- milking** --|	shearing					|lambs born| slaughter for winter*	
GOATS	|---- year-round grazing in pastures, fields, and forests around the village, weather permitting (or may spend winter in barn) ----|										
MUSHROOM COLLECTION						|--- primarily chanterelle and king boletus ---|					
FIREWOOD COLLECTION						|--- firewood cutting and transport to village, most intense August to October ---|					
WILD PLANT, HERB, SNAIL AND FRUIT COLLECTION					-cowslip -skripalets (wild green)	-snails -pine shoots -other herbs for tea and seasonings (e.g., thyme, oregano, yarrow)	-elderflowers -St. John's wort -strawberries -raspberries -blueberries		-rosehips -sour apples -hazelnuts -walnuts -wild plums -cornel cherries			
HUNTING	*(Animal pests such as foxes and wolves may be hunted year-round.)*										|- -legal hunting season for boar - -| *(animals and season vary by year)*	

Notes: Timing shown is for common occurrences of activities. They may also take place at other times. For example, someone was collecting hay in November. "Weather permitting" usually refers to snow- and ice-free conditions, in some cases combined with no precipitation of any kind and/or mild temperatures.

* Animals may be slaughtered at other times for special celebrations (e.g., Easter, Bairam, graduations, village festivals) or for immediate consumption.

** Animals are milked after the lambs have been sold for Bairam and Easter. The dates vary by year.

Figure 5. Seasonal organization of major resource-related activities in Zaburdo during 1997

continued milk production. (On rainy days, of course, hay cannot be harvested, but mushroom collection can be a profitable activity.) With the cooperative cow-herding system, household members need only spend days as cow herders periodically, which allows them the time to take care of crop production, and feeding and milking can be scheduled in with other activities. (Livestock care does require someone available in the village every day for these chores.) For households with many sheep, taking the animals to pasture every day can tie up a household member for the summer, plus possibly other times of the year when snow conditions permit the sheep to be taken out. If necessary, however, sheep can be sent to a shepherd in another village for the busy summer season or kept in the barn for the winter. Most village households are able to organize the labor necessary to carry out these resource-based activities, except perhaps for the less critical ones such as picking raspberries or blueberries, which may slip through the cracks. Summer is seen as a time of unending hard work, from dawn to dusk or beyond, while winter is a time of rest.

But postsocialist village life is not only a matter of natural resource–based activities, and thus labor-organization conflict can occur at another level. Specifically, this conflict is between nonagricultural jobs and resource-based activities, such as working in agriculture, herding livestock, collecting mushrooms, or collecting firewood. Although pay may be low, a job can provide cash income throughout the year and the potential to eventually receive a pension. On the other hand, collecting mushrooms may generate substantially more money for a day's work. Similarly, harvesting potatoes or getting the hay in before it is rained upon can be important for both generating cash through product sales and provisioning the household for the winter. Although some villagers complain that they cannot help everyone—that their beans froze while they were working for others, that their hay got wet because a husband had to work and was unable to transport it to the barn in time, and so forth—most are usually able to deal with this labor crunch in numerous ways, including taking vacation days or doing agricultural work when they get home from their paying jobs. Since most jobs are in the village, many employers understand the periodic need to take time off for agricultural activities, and annual leaves can be generous by U.S. standards. While forestry employees are more likely to have paid work in the summer, the opposite is the case for school employees, since summer vacation generally coincides with the busiest period for agricultural activities. Through these strategies, most households seem able to get their basic work done—albeit sometimes by working for weeks without a break and with some compromises about what is done and when. In the end, my impression is that it is not necessarily the people who are employed who are the worst off, but rather those with limited cash income, few *blizki* in the village, and limited land and household labor (recall the Marinovi household described in chapter 3). They are the ones who must put off their own work to earn some cash and are hard pressed to hire people to cut their hay or pasture their cows. But to flesh out such cases, it is best to turn to additional examples.

Some Examples of Household Economies

Several examples from village households illustrate the ways in which they make ends meet in the postsocialist period through a combination of resource-based activities, nonagricultural pursuits, and other sources of income. Specifically, households are discussed in terms of exploitation of land and other productive resources, labor resources, household labor allocation, and the economic activities in which they participate. Income figures are self-reported and refer to conditions at the end of 1997. Similarly, yield data are self-reported estimates (I have converted the unit of measurement from gunnysacks to kilograms). The cases that follow provide a portrait of how households are surviving postsocialism and the role of resource exploitation in this survival.

CASE 1

Mariana Hadjieva is a 60-year-old widow living alone. She and her husband lived and worked in town for many years, returning to the village in the early 1990s after their retirement. Her only child, a grown son, still lives in town with his family. Though Mariana has more than enough land and expects to farm as long as possible, her activities are somewhat limited by her age. She grows potatoes, beans, and onions on 1 decare of land and mows 3 decares of hay to provide food for her cow and calf. She rents out several additional decares of land in exchange for assistance with agricultural labor and also goes to work for other people, who then return the days, and thus she avoids paying money for help with her agricultural activities. She most often exchanges agricultural labor with her sister and nieces, and her son comes from town twice a year to help with harvesting potatoes and cutting hay. Mariana says, however, that as a town dweller, he is not accustomed to the hard physical labor of village life and finds these activities tiring. Her daughter-in-law, who is from another part of Bulgaria, consequently recommended that she replace the cow with chickens in order to render the hay production unnecessary. Mariana rejected this idea, however, saying that chickens do not give milk like her cow does, and that chickens also must have a place to live and someone to feed them. (Although I did not speak to this son, other sons and grandsons of villagers, who visit Zaburdo to help with agriculture, express concern about the cost of the trips and thus how long they would be able to afford this travel.) Male relatives in the village help Mariana with plowing and transport the hay and potatoes in their truck, sometimes refusing to allow her to pay them for the service. Even so, she said getting her potato fields plowed was particularly worrisome. She also pays someone, in this case a relative, to take her turn with the cow herd.

Like other households, Mariana's livelihood resources come from both formal and informal sectors. She collects a few wild herbs and fruits for personal use and receives a monthly pension of 34,000 to 44,000 leva (about $20–25). Excess milk from her cow is sold to the dairy, for cash or in exchange for cheese, and the calf is slaughtered to provide the household with meat for the winter. Besides keeping potatoes to eat, feeding the culls to her livestock, giving some to her son, and saving seed for the next year, she will exchange "a lot" of potatoes for a long list of food products and animal forage. She prefers the barter system to selling the potatoes for cash. With these sources of cash income and subsistence goods, Mariana is able to acquire food and pay for expenses such as the electric bill. She said, however, that she cannot go on like this forever. She has not purchased new clothes for herself in several years, for example, and eventually her savings will run out. On the day we spoke, she was struggling to decide how much money to give her grandchildren for the holidays—yet another economic obligation. Not only would they expect something for

New Year's Day,[14] she said, but their birthdays were also coming up, and with the return of religious holidays following the collapse of communism, there was also Christmas. Consequently, they would expect to receive gifts on several different occasions in the near future, and the amounts she mentioned added up to nearly the value of her monthly pension.

CASE 2

The Konstantinovi household was one of the largest in my sample, with seven members spanning four generations. The household includes Grandmother Mila in her mid-eighties (the mother of Ivan); the male and female householders, Ivan and Renata, who are in their late fifties; their son and his wife, Stoyan and Marinka, both of whom are in their mid-thirties, and two school-aged grandchildren. Stoyan was a driver during the socialist era and now drives his own truck as a private business. He is the only member of the household regularly employed outside agriculture. His father, Ivan, is a shepherd by profession and now cares for the family's flock in the postsocialist period. Ivan's wife, Renata, worked as a general laborer on the cooperative farm. Now unemployed, she engages in the same activities in the household's private farming effort. The daughter-in-law, Marinka, worked in the electrical assembly workshop until it closed at the end of the socialist era. She now devotes her time to the household's production activities and other chores of domestic life.

With a relatively large pool of labor, this household is heavily involved in both crop production and animal husbandry. They farm 25 decares of land—5 decares in potatoes and 20 decares in hay. Some land is rented for money to produce more hay for their animals, and hay is also purchased. Besides potatoes and hay, their garden contains beans, onions, cabbage, pumpkins, leeks, parsley, mountain spinach (*loboda*), and carrots. During the spring, the household had 50 sheep, 6 lambs, 2 cows, 1 mule, and 4 chickens. By winter the number of sheep had declined to 35 through a combination of sales and consumption, and the number of lambs declined to 4 in response to low prices for meat, milk, and wool. Even so, the Konstantinovi owned one of the larger flocks of sheep in the village.

These production activities combined with Stoyan's nonagricultural job make for a rather complex organization of labor. Ivan is occupied full-time taking care of these sheep plus a few "foreign" ones—for example, those of his brother. He commented that he is not able to participate in the household's other agricultural activities because of this obligation. Grandmother Mila can no longer work, much to her frustration, and the grandchildren help only rarely. This leaves Renata, Stoyan, and Marinka to do much of the agricultural work. Most commonly, the Konstantinovi household exchanges agricultural labor with Ivan's brother and sister-in-law and with Marinka's parents. They also regularly hire people to help with harvesting potatoes, cutting hay, transporting manure, and plowing the fields. The household always pays for another cow herder to fulfill their lot, because Stoyan is employed full-time and Ivan is always with the sheep. Stoyan sometimes must pass up his paid work, and thus the income he could have otherwise earned, in order to finish tasks for the household's agricultural production. All in all, however, this household identified no significant problems in its potato production or animal husbandry activities.

In making ends meet, the Konstantinovi benefit from a diversity of income and other livelihood resources. Household members collect some wild fruits and herbs, but only for household consumption. In contrast to the modest amounts sold by most other households described here, this one sold 10 to 12 tons of potatoes in 1997, and this was after accounting for those kept to eat, those destined to be fed to the animals, those saved for seed, and those given to friends in town. Because the Konstantinovi own a large truck, they have the option of selling their potatoes and animals either to buyers who come to Zaburdo or to consumers in more populous areas. Few potatoes are exchanged directly for goods by this household; however, it does participate in the more informal system, in which members sometimes take milk and potatoes to friends in the lowlands and the friends bring or send them goods produced in those regions, such as tomatoes, cucumbers, and walnuts. Since they have several livestock, they slaughter sheep for meat for the household more often than average. The Konstantinovi regularly sell both sheep's and cow's milk to the dairy, and periodically they sell animals and wool. Grandmother Mila gets a small pension of about 30,000 leva each month (approximately $17), and Ivan is nearing retirement age and thus will soon receive one. Finally, Stoyan brings in cash income from his business.

Given these resources and the variety of activities in which household members are engaged, they are considered to be doing okay to well financially, depending on to whom you speak. Stoyan, for example, described the household as "middle class," while another villager commented that the grandchildren are usually well dressed and are often seen wearing new clothes. The day after selling their potato crop in 1997, Stoyan and his wife returned from town with a new CD stereo, showing some level of discretionary income. Even so, his mother, Renata, is concerned about monthly cash flow. One day, for example, she totaled up their typical monthly income from her mother's pension along with that which her husband would soon be receiving, plus their milk sales. She commented that this amount would suffice for bread and utility bills, and otherwise they could eat what they had in the pantry. While Stoyan is relatively happy with the cards that they have been dealt in the postsocialist period, he and his wife are also unsure about what the future will bring. His mother, on the other hand, would be willing to give up her animals and land to see the return of the cooperative farm and more employment opportunities for village residents, and she says that she would work there every day.

CASE 3

The Ilievi household consists of a young couple in their late twenties, neither of whom is regularly employed outside the household, along with their school-aged son. Elena and Mitko studied tourism-related professions in high school and subsequently returned to Zaburdo. Following in his father's footsteps as a shepherd, Mitko worked in the cooperative farm's sheep-production unit briefly before it was liquidated. Elena worked in the sewing workshop until her maternity leave, but it closed before she went back to work.

As with most village households, the Ilievis raise both potatoes and livestock. Besides about 3 decares of potatoes, they have a garden and grow beans, onions, spinach, carrots, garlic, sugar beets, summer squash, and kohlrabi. Although they exchange agricultural labor with both of their parents' families, they work more frequently with Mitko's family because their fields are at the same location. This household owned six sheep, six lambs, one calf, and one mule in the spring, and the number of sheep had increased to 15 in the fall because the lambs grew up and additional lambs were purchased. Animal husbandry and hay production are done in common with Mitko's brother. The brothers work together in harvesting about 100 decares of hay meadows, some of them rented, and sometimes working with the assistance of paid laborers. They also share the herding responsibilities for their joint flock of sheep.

Almost all the household's income comes from resource-related sources. In 1997 it will sell 2 to 3 tons of potatoes for money, thereby earning about 800,000 to 1.2 million leva ($445 to $665), and it will trade an additional 200 kilograms of potatoes for forage, vegetables, and clothes. In addition, friends and relatives from the Plovdiv area regularly send them goods not produced locally, such as tomatoes, peppers, and cabbage, receiving potatoes in exchange. From the livestock, they sell milk, sheep, lambs, and wool. Both Mitko and Elena collect wild mushrooms for sale. Because the wife is unemployed, she collects mushrooms nearly every day with her sister, while the husband does so less frequently because he is busy with the sheep. According to their estimates, she earned about 500,000 leva ($275) from mushrooms during one summer and he made about 200,000 leva ($110). In the past Elena periodically worked as a day laborer in agriculture to earn money, but this year she collected mushrooms instead because of the higher earning potential. She also collects wild fruits and herbs for household use. The previous year Mitko occasionally earned money by working as a paid cow herder; however, with the increased number of household sheep he no longer has time to do this. This household is doing a reasonable job of making ends meet through their involvement in resource-based activities. Mitko, for example, pointed out that they have been able to build a new house in the 1990s. Under current conditions in Bulgaria, he sees no option other than continued participation in agriculture, although both of them hope that their son will have a career outside agriculture.

CASE 4

Stefka and Petur Todorovi are a retired couple in their late sixties who have lived in Zaburdo all of their lives. Stefka worked at the sewing workshop before her retirement, and Petur was a mid-level manager on the cooperative farm. Their children and grand-children live in town; however, the son, Doncho, also participates in the labor and decision making regarding the household's potato production. Although advancing years and health problems affect the amount of work they can do, the Todorovi are able to mobilize sufficient labor to produce a sizable amount of potatoes on the household's larger-than-average landholdings through networks of kin and *blizki*. Household potato production takes place on about 7 decares of land, and the household also harvests 25 decares of hay. Onions, garlic, beans, pumpkins, kohlrabi, and sugar beets are grown in small gardens. Stefka does a larger share of household agricultural work due to her husband's health problems. The couple works most closely in agricultural production with the parents of Doncho's wife, who also live in Zaburdo. In addition, Doncho and his family often spend part of the summer in the village and help with the agricultural activities. He sells the potatoes in town and also buys the fertilizer, animal forage, and sometimes seed potatoes in larger and lower-elevation communities; in doing so, he overcomes one of the more difficult aspects of the postsocialist system of product distribution to villages. The Todorovi's livestock holdings consist of one cow, two goats, and one horse, and they pay someone to take their lot for the cow herding because of Petur's health problems. The goats stayed in the barn the year I conducted this case study because there was no one to take them out to graze, and they were later eaten for the holidays.

With these activities, plus wild-product collection and pensions, the Todorovi household is financially stable. They are able to purchase presents for their grandchildren, for example, and to pay the transportation costs to visit them periodically. They also have a relatively new Korean television set. After setting aside food for themselves and the children, food for the animals, and seed for the next year, there were perhaps 8 to 9 tons of potatoes left to sell (some of the proceeds will go to the son). The household will also trade about 250 kilograms of potatoes for goods. Stefka collects wild herbs and fruit for household use. The couple does not collect mushrooms for sale, although their son, daughter-in-law, and grandchildren do so when they come to visit. Both Stefka and Petur receive pensions. His is 67,000 leva ($37) a month and hers is 33,000 leva ($18) a month. Friends and relatives in the lowlands send them products including grapes, wine, grape brandy, cherries, and tomatoes. One winter day when I dropped by to visit, the male householder had prepared pumpkin stew, from a recipe he had found in the newspaper, so that his wife did not have to cook when she got home from working on one of their fields, thus showing the flexibility of labor organization in response to constraints on the activities of household members.

CASE 5

Chavdar and Sevda Dimitrovi are a couple in their late forties, and their household also includes son Boris, an unemployed college graduate, and daughter Krassimira, a student at the village school. Chavdar works at the village school, and Sevda was recently laid off. Besides 2 decares of potatoes and 12 decares of hay, the Dimitrovi grow beans, leeks, onions, garlic, summer squash, celery, carrots, parsley, kohlrabi, strawberries, and raspberries. Agricultural labor organization is not a problem, because only Chavdar is employed and his job leaves him with summers free. When necessary, they exchange labor with their siblings in the village. The Dimitrovi's livestock in the spring consisted of two cows, one calf, some rabbits, and a donkey. (The donkey was later exchanged for a horse.)

These agricultural activities, combined with some cash income and collection of wild products, provide the Dimitrovi household with the subsistence and cash resources with which it is surviving postsocialism. Besides retaining potatoes for food and seed, it will sell about 2 tons of potatoes and exchange 100 to 200 kilograms for goods. Milk is sold to the dairy nearly every day. All household members collect mushrooms during the summer, and their combined estimated earnings of about 1.2 million leva ($667) may exceed their earnings from potato sales. They also collect wild fruits and herbs for personal use. In late 1997, Chavdar earned 110,000 leva a month (about $60) from his regular job, and he occasionally did carpentry work on the side. When Sevda was laid off, she took her nine months of unemployment benefits as a lump sum payment of 300,000 leva ($167), because they needed cash at the time.

Given current conditions, neither Chavdar nor Sevda see the possibility of not participating in agricultural production as a viable option. Without these agricultural activities they could not survive on a salary, particularly with only one household member employed. Like several other villagers, they say that things were better under socialism. Now they have enough money for food and other immediate expenses such as utilities, but little else. Under socialism, they had enough money to go on vacation and to buy the children new clothes several times a year. In contrast, the daughter Krassimira has now worn the same jacket for three years, said her mother. The son, Boris, for his part, does not want to live in Zaburdo but will do so if there is no other choice. Although he studied a field related to agriculture in college, he does not want a job in agriculture. He said that it is difficult to find jobs in town under current conditions, and living expenses in town are also high. So, for the moment at least, he is living in the village.

These household case studies, along with those presented in chapter 3, illustrate the ways in which differing combinations of land, labor, and other sources of income are employed to provision village households and make ends meet. They also provide some insight into the relative importance of different activities for generating cash. Some households, particularly those with substantial labor and land resources or the ability to mobilize them from outside the household, are able to do more than just survive. New televisions are purchased, children are sent to college, and gifts are given to grandchildren. This requires considerable hard work by household members, however, particularly during the busy summer season. Others are surviving, but just barely, and these people may feel strongly that life was better under socialism. Older and ill villagers, in particular, are between a rock and a hard place if they must pay for someone to herd their cows or plow their fields out of a small pension.

In all cases illustrated, the households participate in a diversity of activities and income-generating strategies. All of them engage in subsistence-oriented activities and many also produce potatoes, milk, or both for sale. Mushroom collection is another popular source of income, particularly for younger people. Some households also have members employed in the formal economy, who receive pensions or who are involved in business ventures, although given the failure of these income sources to keep pace with inflation in recent years, one cannot rely on them alone. In other words, village households are partially articulated with the modern cash economy but also rely heavily on informal, subsistence-oriented activities for their daily livelihood. Indeed, unemployment of one or more family members sometimes prompts increased resource use in an effort to make up for lost income. This resource use, in turn, is highly dependent on the resources available in the territory around Zaburdo.

Perceptions of Postsocialist Patterns of Income Stratification

Many village residents complained to me that their economic status (and that of their neighbors) had declined during the postsocialist period as a result of both job loss and the devaluation of whatever wages, pensions, and savings they had.[15] Some of them also observed that everyone used to be more or less equal, with no one really rich or really poor, while now there is more inequality (see also Creed 1998: 265–273).[16] On a visit to the village in 2000, I explored with a handful of villagers their understanding of postsocialist economic stratification in Zaburdo.[17] The individuals I spoke with included people who are retired or engaged primarily in farming (although they would not necessarily identify themselves as farmers), as well as people employed at the school, the local government administration, and the health service, who were thus in positions to examine the question from a professional point of view. On the whole, the answers I got were remarkably consistent, and they also more or less corresponded with my own observations.

Most people described a three-way system—rich, middle or average, and poor—and money seemed to be the primary criterion for categorizing households. Material possessions were often mentioned as well; however, some people I spoke with cautioned that one cannot necessarily judge wealth based simply on the presence or absence of goods such as cars and television sets. These items might have been purchased at a time when the household was better off, and in any case, most village houses are relatively similar, and most people have TVs. One person said that the number of rich people was stable, while the number of poor was increasing.

While it was fairly clear who among the households was very poor, it was more difficult to identify a dividing line between poor and average. Indeed, one person simply said that there were a handful of rich people and that the rest were poor, declining to differentiate between poor and average. Most others I spoke with had suggestions about how to identify poor households, however. The poorest people are eager to work for others for day wages, even if it means neglecting their own agricultural tasks. They have nothing to eat and no regular income, no pensions, and no jobs. They buy bread on credit and pay when they get money from their milk sales. At school, the teachers notice that their children's shoes that are falling apart and that for lunch, instead of a packaged croissant or wafer, the parents send their children off with a piece of bread spread with pepper relish. The poorest usually have no television or refrigerator, or if they have an old one, they cannot afford to get it repaired.

Many people I spoke with said that most village residents are average economically, although a few clarified this evaluation by saying it was made according to village standards.[18] Those who are "average" are able to get by from month to month without regularly needing loans or credit, but they also have no money for luxuries. They may own a house, livestock, a barn, a Western TV, and possibly a car but they have not vacationed recently. During village festivals they are out eating and drinking, but not to the degree seen with the rich. Pensioners may be included in this middle category, as pensions provide regular income, although a lone, often female pensioner with a minimum pension, and lacking family or *blizki* to help with agriculture and provide firewood would fall into the poor category.

Everyone agreed that there was only a handful of rich people in the village, often naming them by name and usually identifying the same names. (Such naming rarely happened with the other two categories, except for a few people who either classified themselves or named the poorest family or two in the village.) Rich households seemed relatively easy to identify. They had newer Western cars, multiple vehicles, or both; they owned an apartment or other property in town; they had savings in the bank; their children were well dressed; their houses were particularly well furnished with fancy tile floors and woodwork; and finally, they might be involved in trade and perhaps also the large-

scale production of potatoes for sale, for which they hired workers. In the last case, they also had the ability to take the produce to
market, thereby avoiding the middlemen who come to the village.

Postsocialist elites, and how they got to be elites, have attracted
some attention in the last decade. The most common explanations
are that these people have parlayed socialist-era connections and
resources into elevated status in the postsocialist era, or alternatively that presocialist or other factors are more important
(e.g., Kostova 2000; Róna-Tas and Böröcz 2000; Sikor 2001; Słomczyński and Shabad 1997; Szelényi et al. 1998; Szelényi and
Szelényi 1995; Yoder 1999). I did not get the sense from my conversations that one could necessarily predict a household's or
family's current economic status based on the socialist or presocialist past, however. One family that was identified as being
among the better off families in the postsocialist period was reportedly poor in the socialist period, but it was also among the
better-off families during private times, both economically and
politically. Others in leadership positions during socialism continue to do well in the postsocialist period. But in one such example, the grandfather was mayor before collectivization, his son
was a late-socialist and early-postsocialist mayor, and now both
the grandson and the husband of the granddaughter are among
the more successful village "businessmen." And good socialist
connections do not necessarily qualify one for a bank loan, as loan
criteria have been tightened following bank failures, creating difficulties for some families that want to start a business, buy farm
equipment, or pay university tuition as a way to get ahead. Thus,
my anecdotal data on the subject are inconclusive, other than to
suggest that the situation is more complex than many analysts ad-
mit (see also Lampland 2002).

Making Ends Meet

This chapter has examined the ways in which Zaburdo residents
are provisioning their households and otherwise make ends meet
in the postsocialist period by looking at the sum total of activities
in which they engage. Resource-based activities play a particularly important role in providing villagers with food and heat as
well as cash income—perhaps more so than during the socialist

period. Yet, many households also rely on other sources of income, including wages from regular jobs, pensions, and earnings from business or handicraft production, and these other sources are important because they often come in the form of cash. In this postsocialist economy, barter exists alongside sale of potatoes, wage labor alongside mutual labor exchange, and regular employment alongside subsistence production, with each of these modes of exchange and resource generation having particular benefits and drawbacks. From the standpoint of labor scheduling, summer is the busiest season, and the extent of villager involvement in agropastoral production can be seen in the relatively limited number of people who are interested in doing casual agricultural labor in exchange for day wages. Yet, several means are available for coordinating agricultural and nonagricultural labor.

The data presented in the last two chapters show that Zaburdo residents employ multiple livelihood strategies during the postsocialist period. Without the benefit of a historical perspective, one might be tempted to analyze this diversification itself as a response to postsocialist conditions. Yet, similar diversity was seen in the economic activities of local residents both before and during the socialist period. These villagers have long relied on a combination of agriculture, pastoralism, and forest-product exploitation. Wage labor and pensions were added to the economic portfolio, mostly during the socialist era, but many people continued at some level with earlier, resource-based activities for household self-provisioning. Private enterprise is again possible in the postsocialist period. With the recent restitution of formerly private forests, forest exploitation may return as an income source for some. Such diversification involving agriculture, pastoralism, and off-farm resource exploitation, sometimes with the addition of tourism, trade, or labor migration, is common in mountain settings (e.g., Netting 1981; Rhoades and Thompson 1975; Stevens 1993). Hussein and Nelson (1998) suggest that most rural producers have historically diversified their activities to some degree. Two recent reviews of household livelihood strategies observe that there can be a range of motivations for diversification, including seasonality, risk avoidance, coping with crisis or distress, adapting to long-term income declines due to environmental or economic change beyond local control, or even accumulation of money or other resources (Ellis 1998; Hussein and Nelson 1998).

More interesting than the simple observation that most households obtain livelihood resources from multiple sources is a consideration of the changes in their livelihood portfolios over time. While some things have been added or dropped, often the issue is more one of the relative importance of various activities, and perhaps the organization of production or exploitation, rather than simply considering the basic activities in which people engage. In many cases, the changes in activity or the emphasis that an activity receives are at least partially a function of forces beyond the local environment. An early example of this is the reduction in livestock numbers following loss of access to the Aegean pastures. This reduction—along with improved transportation infrastructure and the secured ownership of forests—may have increased the importance of commercial timbering for some households. Changes are likewise occurring in postsocialist Zaburdo, one of which is a partial retreat from the market and wage labor. Along with this change comes increasing reliance on activities that are subsistence-oriented and dependent on local natural resources. Few jobs are available locally, and jobs in town may not pay enough to cover living expenses for food and lodging due to the general economic situation in the country. Agricultural production is similarly made difficult by current conditions due to high input prices, lack of mechanization, and problems with product marketing. Verdery (1999) raises concerns about the extent to which postsocialist landowners have the means for using their property, and Zbierski-Salameh (1999) likewise observes that Polish peasants in the 1990s were retreating from the market and returning to more subsistence production, in response to a context in which access to mechanization, commercial inputs, and the means for disposing of production are difficult. Of course, such a response is not necessarily unique to postsocialist Eastern Europe. Cole and Wolf (1974:89) similarly write of a return to subsistence production in the Italian Tyrol between the two world wars as the marketing of produce became unprofitable. Meanwhile, the collection of wild products, for which a system exists for marketing the produce, has perhaps increased as a means for generating cash. Here, too, is clear evidence of the influence of external political and economic conditions on resource use.

The level of resource exploitation by residents of postsocialist Zaburdo is clearly related to larger political and economic conditions and their inability to survive on wages or pensions alone. Similarly, the use of agricultural inputs and the types of technology employed are constrained by economic conditions as well as ecological ones. (Commercial credit is difficult to obtain because agricultural land and village houses are worth little if anything as collateral, and agricultural-specific development funds are often not interested in supporting the scale of production of most village households.) Meanwhile, many young people I spoke with are not interested in jobs in agriculture; they want to move to town and have professional careers. This, too, has important implications for the future of resource use and management in Zaburdo and other Rhodope villages.

The Rhodope Mountains—the land of Orpheus. The only things visible from ridge tops are pastures, thick evergreen forests, pastures and mountains. On a clear day, it is said, you can see the Aegean Sea.

The "Wonderful Bridges" (*Chudnite Mostove*) are natural rock arches located about two hours walk from the village. Nearby are two mountain chalets, where tourists sometimes stay while visiting the area.

The hay meadows and potato fields of village farmers. At closer range, one sees the hay meadows and potato fields of village farmers, their borders sometimes discernible as lines of brush or low stone walls. The buildings clustered in the foreground, center, date to the presocialist era, when families often had outbuildings located near their agricultural holdings.

Outbuildings used by a family with a flock of sheep. Although outbuildings are often unused and in poor repair, this particular cluster is currently used by a family with a large flock of sheep. On the day of my visit, some additional villagers had arrived to help replace the roof on one of the buildings.

A view of the village from one side of the valley. Due to the rocky terrain, houses are perched on the steep hillside. Many are built of stone or brick, and their roofs are often covered with red ceramic tiles or stone slabs.

Lining up for the bus. Four days a week, a bus leaves the village early in the morning (except on Sunday, when it leaves late in the morning) and returns in the evening. Although service is significantly reduced from that of the socialist era, this bus remains an important transportation link for villagers who do not own cars. On Friday afternoons and just before lunchtime on Sundays, the bus is often filled with teenagers who have completed the village school and are attending secondary school elsewhere.

Waiting in line to buy bread. Bread is not in short supply; rather, these men, women, and even a few children are simply waiting for the next batch to come out of the oven. The bakery is still run by the local commercial cooperative, as it was during the socialist era.

Elderly and poor villagers occasionally receive financial or in-kind assistance from the Bulgarian government or external donors. In this photo, pensioners fill gunnysacks with coal that has been brought to the village to assist them during the winter heating season. Although the coal itself was provided free of charge, the recipients reportedly had to pay part of the transportation costs.

Agricultural supply sales on the village square. The village lacks a regular store at which residents could purchase agricultural supplies such as fertilizer or animal feed. Consequently, a common sight is one or more trucks in the village square selling 50-kilogram sacks of these items. Sometimes the sales take place using money; in other cases, villagers barter for these products with potatoes that they have grown.

Preparing *korban*. These men are preparing a thick stew of mutton and grain for a ceremonial meal called *korban*. Villagers sometimes gather in the square to eat the stew together, but on this day they simply brought containers in which to take the food home. While celebrating a religious holiday is one occasion for such meals, for this occasion someone had simply donated money so that the meal could be prepared for the health and luck of village residents.

Egg battles are a popular Bulgarian Easter tradition. Here the author and a friend knock together dyed, hard-boiled eggs to see whose is the strongest. On this occasion my egg was declared the champion, being the last to remain unscathed by the battle. Although historically this is a Muslim village, some residents report that they feel closer to Christianity or otherwise celebrate Christian holidays.

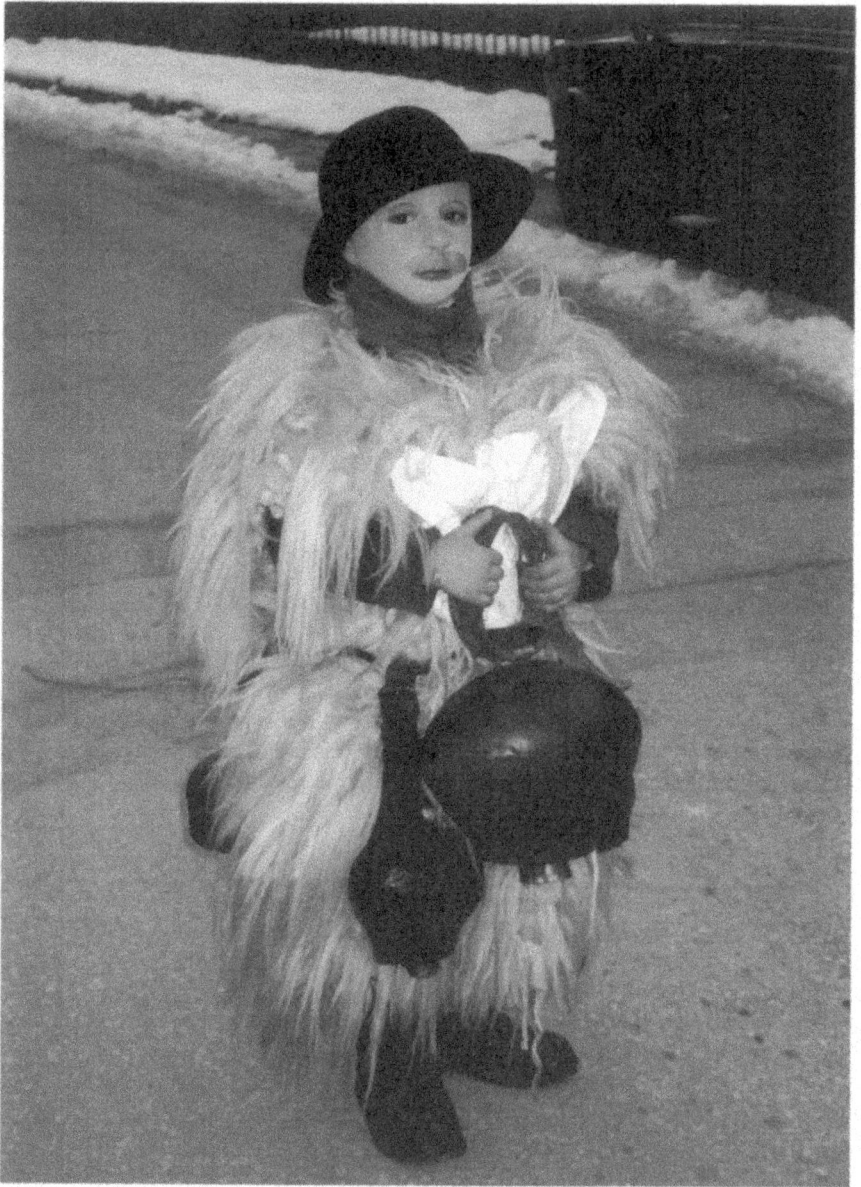

The littlest Kuker. Another popular village tradition is one held on the 14th of March, during which boys and young men dress in masks or costumes made of sheep or goat skins and decorated with sheep bells, the bigger the better. According to local lore, they go from house to house in the early hours of the morning, scaring away the evil things like snakes and lizards that come out when the ground warms up in the spring. In exchange for this service, the costumed participants collect token gifts from the householders—in an earlier time they received eggs, but now money.

Preparing for the potato-planting campaign. Small seed potatoes saved from the previous year's crop are put in wooden crates and left in a warm place for a week or two until they begin to sprout. At this point it is time to transport them to the fields for planting. Here a retired couple and their teenage grandson prepare to load the seed potatoes into a truck for transport.

Making repairs on a pack saddle. Although there is no saddle maker in village, some villagers are skilled at repairing saddles. Straw is sometimes required for the repair work.

Waiting for the cows to come home. In this typical evening scene, the neighborhood cow herd returns from a day in the pasture to waiting villagers, who milk and feed them.

Transporting milk to the buying point. For households with cows, one morning chore is transporting milk to the buying point. Here two neighbors cart their empty milk cans home.

A small private sawmill. Besides agriculture, timber production is an important economic activity and use of natural resources in this region. This young man hoped to cash in on the forest privatization by providing timber processing, although later he realized that he would not make any money and sold his machinery.

Spit-roasted yearling lamb or mutton is a favorite regional dish. Here two men from the village prepare it for sale at a regional folklore fair held at Rozhen. The meat was then sold at the restaurant of the grandson of the man turning the spit.

The girls singing group from the village also performed at the Rozhen folklore fair. The singing group is affiliated with the village *chitalishte* and directed by the *chitalishte* librarian.

"Support the establishment of Nature Park Western Rhodope." So says the sign on the only booth at the Rozhen folklore fair to have an NGO presence. The booth was staffed by members of two Bulgarian environmental organizations— Green Balkans and the Bulgarian Society for the Conservation of the Rhodope Mountains (BSCRM). On the sign one can also see the logos of the organizations, a chestnut leaf for the Green Balkans and a sheep bell superimposed on a pine tree for the BSCRM.

5

Conserving the Natural
Heritage of the Rhodope

With the fall of state socialism in Eastern Europe and increasing
interest in and support for conservation in the region, the
Rhodope Mountains have attracted the attention of NGOs and
government bodies, both Bulgarian and foreign. Efforts in the
1990s to create a large-scale protected area in the Rhodope were
hampered by concerns over property ownership, because of Bulgaria's postsocialist restitutions of agricultural land and forests,
and also by the country's lack of a modern nature-conservation
law for much of the decade. Yet, several NGOs have engaged in
conservation-related activities in the region over the last decade,
and additional efforts are planned. Following up on chapter 1's
general discussion of NGOs and nature conservation in Bulgaria,
this chapter focuses on NGOs as they relate to rural areas that are
of concern to conservationists. Although the primary focus is on
the Rhodope, I occasionally draw on experiences elsewhere in
Bulgaria, particularly concerning relationships between Bulgarian NGOs at various levels and international donors. Case studies of particular organizations, individuals, and activities provide
insight into the relationships between these Bulgarian NGOs and
international conservation organizations, on the one hand, and
between the Bulgarian groups and rural Rhodope communities,
on the other. In so doing, this chapter addresses the question of
how and to what extent environmental organizations are directly
affecting local-level resource use and resource management in
the Rhodope.

This chapter is organized into three sections. An analysis of the spatial dimensions of the NGO community in the first section examines where environmental groups in Bulgaria are located and the implications of this geography for the issues they address and their connections both to rural communities and to international donors. The second section focuses on the role of individual leaders in Bulgarian environmental organizations and especially the kinds of people involved in NGOs in rural settings. The third section addresses in more detail the connections between Bulgarian and international organizations, on the one hand, and between Bulgarian groups and rural communities, on the other. In so doing, it also describes some conservation-related activities by NGOs, particularly in the Rhodope, and the way in which the activities of specific organizations are affecting particular communities at particular points in time. Taken together, the discussions in this chapter further illustrate the characteristics of the Bulgarian environmental community and how they affect the relationships between the global and local arenas with respect to biodiversity conservation and natural resource management.

On the Importance of Place: A Geography of Bulgarian Environmental NGOs

The geography of environmental organizations is an important characteristic to consider in mapping the conservation landscape, particularly for its potential to influence relationships between Bulgarian NGOs and rural communities, on the one hand, and between Bulgarian NGOs and the Bulgarian government or external donors, on the other.[1] Such a discussion also raises issues of organizational structure. More than half the environmental NGOs listed in the 1997 *Catalogue of Environmental Nongovernmental Organizations in Bulgaria* maintain addresses in the five largest Bulgarian cities—Sofia, Plovdiv, Varna, Burgas, and Ruse—and nearly 45 percent of them are in the capital city of Sofia alone (Penchovska et al. 1997). To put this in perspective relative to the country's population distribution, about 13 percent of Bulgarian residents live in Sofia, and 25 percent live in the largest five cities combined (National Statistical Institute 1996a:41, 44). While such a tendency for urban concentration is not unique, Bulgaria, along

with Albania, has significantly fewer environmental NGOs located in areas outside the capital city compared to other countries
in Eastern Europe (REC 1997:15). Thus, one might say that there
is an urban bias to environmental NGOs in Bulgaria, but the more
important question is what this means for NGO activities, priorities, and connections.

In discussing Albania, Sampson (1996) observes the tendency
of NGOs to be located in the capital and says that this can be an
advantage for contacts with foreign donor organizations. Wedel
(1998a) makes a similar point about an urban bias in international
financial support to postsocialist countries. While there is some
truth to this, the issue of location could create other dilemmas for
conservation organizations and other groups dealing with rural
issues. Being located in the capital can be important for the proximity it affords to foreign contacts, national-level decision makers, and decision making. In some ways, this urban concentration
is not particularly surprising, since environmental NGOs in Bulgaria are often composed of environmental professionals and students in environmental fields, and the highest concentrations of
such individuals are found in urban centers with research institutes and universities. But, such a location puts the NGOs farther
away from the places that they work to protect and the people
who live there.

Although Bulgaria is a small country and until recently was
relatively well served with bus and train service, it still takes several hours to get from the capital city of Sofia to Smolyan in the
central Rhodope Mountains or Burgas on the Black Sea Coast,
and additional time is needed to travel from there to smaller
settlements. Relying on public transportation for visits to rural
villages to talk about sustainable resource use or conservation
efforts can be a challenge. Few Rhodope communities can be
reached by train, the frequency of bus and train service has been
reduced in the postsocialist period, and the bus system is largely
designed to serve village residents rather than visitors. (Buses
leave villages early in the morning and return in the evening, and
they do not necessarily operate every day of the week.) An associated question is how to pay for transportation costs (a bus or
train ticket, or gasoline if a private vehicle is used), given the limited resources of many organizations and their reliance on project
funding from foreign donors. Recall from chapter 1 that a rural

tourism organization was unable to travel at one point because its gas money had been spent on office paper. On another occasion the same organization was able to travel courtesy of a foreign visitor who paid for the gasoline. Thus, Bulgaria's transportation network does not necessarily assist the NGOs in dealing with the problem of where to locate themselves. This is seen, for example, in the fact that one reason the youth section of the Bulgarian Society for the Conservation of the Rhodope Mountains (BSCRM) established itself as an independent group was due to problems with activity coordination between Sofia, where the youth section leader lived, and Chepelare, where the larger society was based.

Some NGOs are aware of this geography and what it means for their relationships with rural communities, external donors, and government policymakers. The issue of location was raised, for example, during the selection of a new president for the Bulgarian Society for the Conservation of the Rhodope Mountains after its president's unexpected death in 1997. As a preface to the election, it was suggested that the ideal president would frequently be in both the Rhodope and Sofia, have a driver's license to facilitate travel between the two places and within the Rhodope, speak a Western language, have time to be president, have computer skills, and know the organization inside and out. This was described as an ideal to which the organization should aspire, although the person making the suggestion said that it might not be possible to find someone with all the enumerated characteristics. This list clearly shows concern with linking the capital city and the mountains as well as concern about maintaining foreign contacts through knowledge of a Western language.

After hearing this discussion, I raised the issue of location with officials in the Plovdiv-based Green Balkans. According to one official, being situated in Plovdiv gives the organization's activists a better perspective on life, since one can lose touch in Sofia. At the same time, he said, the organization was closely tied to the capital so that not being present there was not a handicap. The group has good friends and colleagues in Sofia who are frequently in the Green Balkans' Plovdiv office (and vice versa). There is a Green Balkans branch in the capital, and frequent public transportation links the two cities.[2] Thus, this respondent's answer essentially confirms the importance of being in Sofia for the connections that it affords, while at the same time indicating a value in being outside the capital for the perspective this provides. Recall here, as well, the concern of this organization that moving its office would disrupt its connections with foreign contacts.

The examples above are from established organizations with relatively strong ties to the capital and other urban communities. The contrast in terms of geography becomes even more clear when the focus moves from a city to rural Bulgaria. The examples that follow illustrate how the approaches taken and issues addressed by city-based organizations can be fundamentally different from those of NGOs rooted in the countryside, and these differences stem from diverse sets of priorities and different perspectives on what is important. This divide is also seen in the comment of a local government official in one rural community, who said there are no environmental organizations in that small town because there are no pollution problems and thus no need for one. In some ways this issue returns to the question posed earlier about how to define an organization as an environmental one. Environmental issues, at least as the urban environmental community typically knows them, may not be at the top of the agendas of the local groups. For them, other things may be more important. Foundation "Orpheus-Draginovo," located in the western Rhodope village of Draginovo, is concerned about sustainable development—but sustainable development of the community, through activities such as providing adequate drinking-water supplies and preserving cultural traditions, rather than specifically sustainable development of the environment. Similarly, the first externally funded project of the Center for the Sustainable Development of the Mountains–Smolyan, a small organization established in the central Rhodope town of Smolyan in the mid-1990s, seeks to address the region's unemployment problems by promoting environmentally friendly, family-owned businesses. Other project proposals under development by the center concern health care and agricultural development. Thus the interpretation of what constitutes sustainable development or even an environmental issue can depend upon one's perspective. And in the cases described above, an important variable affecting perspective is urban versus rural location.

Leaving the Rhodope for the moment, such differences in perspectives and priorities are also seen in the different responses to conservation dilemmas in the mountains of southwest Bulgaria during the 1990s by rural communities and community-based groups, on the one hand, and the urban-based conservation organizations, on the other. One example comes from the discourses used by several organizations in the struggle against the diversion of water from the Rila Mountains to a reservoir that supplies drinking water for Sofia. Although the ultimate objective of the groups was the same—to prevent construction of the water diversion infrastructure—the arguments were presented in rather different terms. The urban-based groups, such as Green Balkans and Green Patrols, focused on biodiversity issues, such as damage to vegetation and wildlife and the potential to upset the region's water balance. The Citizens Initiative Committee for the Protection of Rila Waters, a group whose members live in the villages near the diversion and which had widespread support in these villages beyond its formal membership, raised issues of local autonomy and their rights to have access to the water (Staddon 1996, 1998). In another case, several Bulgarian environmental NGOs protested when the Ministry of the Environment and the Committee on Forests approved a proposal to expand ski runs in the Pirin Mountains near the town of Bansko—an action that reportedly received a predominantly negative reaction from the local population, who were interested in the economic development benefits of expanding the ski area (Bulgarian Society for the Protection of Birds 1996).[3] These examples illustrate the potential for conflict between the priorities and interests of rural and urban groups, particularly in terms of the relative emphasis put on environmental conservation versus economic development (see Staddon and Cellarius 2002).

In light of this discussion, the Bulgarian Society for the Conservation of the Rhodope Mountains presents an interesting case, because in a sense it has one foot in the city and the other in the countryside. Although the society's primary focus is conservation, its approach to the issue stems from the idea that nature and culture combine to make the Rhodope region what it is and that it is not possible to have one without the other. This concern— which is more explicitly focused on local people than many of the city-based organizations—is also reflected in the organization's logo, which combines a pine tree and a traditional Rhodope sheep bell. (The logos of most other conservation organizations discussed here consist of natural elements—a chestnut leaf, a bird, a tree, a mountain landscape.) Concern for local people is also seen in the way in which some of the society's activities aim to bring money to the mountains to help better the lives of the people who live there. For example, a portion of the funds raised through the society's participation in ecological tourism are given to the communities hosting the visitors, based on the assumption that if local people see that they are benefiting from the local environment they are more likely to take care of it. Indeed, the society's president suggested to me in the mid-1990s that advocating for a protected-areas policy on the national level, and in particular the creation of a park in the Rhodope, was not necessarily the most effective way to protect the landscape. He said that perhaps a park would be created in these mountains sometime in the future, but that in the meantime many things could be destroyed. For this reason, he explained, the organization tries to accomplish things on the ground at the local level and puts less effort into working with the Ministry of the Environment. Recall, too, the discussion of desired qualities of a new president for the organization concerning location and mobility, that is, for this individual to be frequently in both the city and the countryside.

The structure and membership profile of the society are relevant to this discussion of geography and connections. The Bulgarian Society for the Conservation of the Rhodope Mountains has branches in urban Sofia and Plovdiv as well as in several rural villages and towns in the Rhodope. Its founding president and some board members are university-trained professionals who are often in environmental fields—as is frequently the case with Bulgarian environmental NGOs—but they also have close personal or family ties to this rural mountain region. The society's president explained that these individuals assist the society in its work at the national level through their contacts and technical expertise. Similarly, a leader of BSCRM's Plovdiv branch described this urban-based group as being part of an environmental alarm system. People from branches in the mountains can call someone in Plovdiv when they discover an environmental problem or threat, and the members of the Plovdiv branch can then contact the appropriate parties to raise alarm, address the problem, and or take other action.

The Role of Individuals in Bulgarian Environmental NGOs

When considering participation in Bulgarian environmental organizations, particularly in rural settings, a good place to start is
the importance of individuals in sustaining such organizations.
The small size of Bulgarian environmental NGOs and some of the
reasons for this condition were discussed in chapter 1. And the
smaller the community, the smaller the organization. This small
size means that a few key individuals can be critical to organizational stability and sustainability. Jancar-Webster observes the
following about environmental NGOs in Eastern Europe generally: "Frequently the leader is a forceful, assertive individual who
may carry the group as long as he is interested, but when his interest flags, the group dissolves. In every country, the fate of an
environmental group often depends on the initiative and energy
of a few individuals" (1993:208; see also Bridger [1998:214] on
women's organizations in Russia). In the Bulgarian case, NGO
bylaws typically specify the positions of officers, election procedures, and how often elections should be held. Not infrequently,
however, the same person has been president for much or all of
the short life of the organization, and this individual and perhaps
a few others often are quite critical to the group. Consequently,
when a key person or persons moves or otherwise has his or her
attention diverted from the organization, the group may cease to
function. I encountered a couple of cases in which the disruption
was a temporary situation, and the organization subsequently experienced a new lease on life under new leadership, but in other
instances the situation could be permanent. Even for the larger
groups, this loss of key people can be a problem for their local
branches, particularly when the organizational structures extend
beyond the larger population centers and are consequently based
on a few core activists and activities. Such situations were related
to me regarding both the Bulgarian Society for the Protection of
Birds and the Bulgarian Society for the Conservation of the
Rhodope Mountains. The local branches in two small communities disappeared when one key leader in each community moved
to another location or otherwise was unable to continue with the
activity.

   The importance of individuals to organizations is seen in other
ways as well. Specifically, organizations are sometimes associated

with particular individuals, and the social or professional position of the individual leader can help the organization and their
place in it—what one might call the personification of environmental NGOs. The tendency of urban NGOs to consist of environmental professionals or students in environmental fields was
discussed in chapter 1, along with some of the implications of this
tendency. At least as interesting, however, is the question of who
is involved in such organizations in villages and small towns.
Staddon (1996:244, 247), for example, discusses the role of specific
individuals in the Citizens Initiative Committee for the Protection
of Rila Waters. He suggests that the presence of certain leaders
was critical to maintaining community support for blockades set
up in 1994–1995 to protest against the construction of a water diversion near the small community of Sapareva Banya in the Rila
Mountains. He also observes that one of the committee's leaders
was a local school principal and that this professional position
was important for establishing her credibility as a spokesperson
and leader for the organization. Other committee leaders also occupied important social positions in the community, for example,
the director of the *chitalishte* and a local entrepreneur (Staddon
1998:364). Similarly, Koulov (1998:211) explains that the chairman
of the NGO Ecoglasnost Independent Union was also the director
of ecology for the Burgas municipality, and three members of the
organization were on the municipal council. These connections
were noted in chapter 1 in a more ambivalent sense relating to the
intertwining of politics and nongovernmental organizations. Yet,
he suggests that these government-sector positions of the organization's leaders enhanced the group's image among the general
public, because most people saw executive power as critically
important in achieving environmental protection. In some cases,
the organizations themselves are aware of this. The president of
the Center for the Sustainable Development of the Mountains–
Smolyan explained that although it was a young organization, its
members are well-known civic and political leaders in the community—for example, a doctor, a lawyer, and a local elected official. This enhances the status of the organization, and it consequently does not have to prove itself.

This importance of and identification with particular individuals is also seen in the external media coverage of these groups.
One example—from an urban setting but related to conservation

policy—is an editorial that appeared in the Sofia-based newspaper 24 *Chasa* on December 18, 1997, with the headline, "Amadeus, protect Vitosha!" The editorial discusses what it sees as a threat posed by proposed protected-areas legislation to the status of two national parks, including Vitosha Mountain, which towers above Sofia and is described in the editorial as "the lungs of the capital."[4] It makes no reference to environmental organizations by name, instead calling upon Amadeus to "ready the battalion [troops]" and also for "Edvin to return." Amadeus is undoubtedly Amadeus Krastev, the flamboyant national coordinator of an organization called the Green Patrols. The quasi-military structure of the organization—also suggested by the language of the editorial—is reflected in its reference to its subdivisions as "detachments." Edvin Sugarev is a one-time president of Ecoglasnost–National Movement, one of the first independent organizations to register following the fall of the Zhivkov regime and a founding member of the Union of Democratic Forces. Sugarev is perhaps the best-known member of this group and was also a member of the National Assembly for a time (Staddon 1996:263–264). While such connections might be looked on critically by those interested in NGOs as entities separate from the government, these examples illustrate how individuals and their positions can enhance the status of the organization, particularly in a context in which the legitimacy of NGOs is not always recognized or established.

Beyond the question of how individuals are associated with or can enhance the reputation of an organization is the more basic one of who creates and participates in rural NGOs or rural branches of larger organizations. As mentioned earlier, these organizations are usually small and rely on a few key individuals to keep them going. Rural settings also generally lack the more dense concentrations of environmental professionals and students found at urban-based research institutes and universities and who often take on leadership roles in urban-based NGOs. Although activists from several NGOs mentioned Zaburdo to me as a place I might consider during my search for a field site, only one person in one household included in my household survey reported membership in an environmental NGO, and this was through connections in town.[5] This leads me to the question: Who is active in creating and sustaining these organizations? Examination of three examples from the Rhodope help to address this question, and the cases also illustrate more generally the circumstances surrounding the creation of these groups and the connections between the local groups and regional or national organizations.

The first example comes from a local development society in a rural Rhodope municipality. The society is based in a small town that developed around natural resource exploitation during the socialist period and now serves as the administrative center for the municipality. This organization was founded with the goal of assisting with the general development of the municipality, and consequently, according to some, it is not strictly an environmental NGO. Indeed, I had a disagreement one day with a Democracy Network project manager about whether this organization was an environmental NGO. He thought not. I disagreed, because it seemed to me that, in a rural area where the economy is based on mining, logging, and agriculture issues of local sustainable development must also include ecotourism and natural resource exploitation, which is to say that this group addresses environmental issues. The society's primary goals are to participate in the development of democracy in the municipality, to promote sustainable development, and to stimulate cooperation and the exchange of information between citizens, economic organizations, and the local administration.

During the initial stages of my research, the president of this local development society was recommended to me by several people as a good person to talk to in this locality—the emphasis always being on the individual, not on the organization. On the surface, the president appeared to be a fairly typical Bulgarian woman living in a small town—in her thirties or early forties, and married with children. Her parents had moved from another part of the country to take jobs in this town during the socialist era, while her husband was from a nearby village. From here her story becomes more unusual, however. In the early 1990s she and her husband decided to start a small business, and she began looking for resources to help with this endeavor. In the process, she discovered that considerable information was available by looking in the right places; she then came in contact with business development courses run by PHARE, and eventually had the opportunity to visit the Philadelphia area on an international business exchange. On this trip she observed that people in the United States do not necessarily wait for the government to do everything. In particular, she described her discovery of a tourist office that was created by some hotel owners themselves, without waiting for the government's help.[6] Upon her return, she began working with the local government in her community and with the regional development association in a nearby town. Eventually the individuals involved decided that it would be more useful having a separate organization in the smaller municipal center rather working as part of a larger group, and so this is what they did—with funding from the Democracy Network. Although other people participate in the organization, including some high-school students in a youth section, this individual is the most obvious participant, and she clearly plays a key role in the organization. This example is also interesting for revealing the intertwining of experience in the business and nonprofit sectors.

Two additional examples of the kinds of people involved in environmental groups in rural settings come from branches of the Bulgarian Society for the Conservation of the Rhodope Mountains. (Another example is presented in chapter 6.) During my 1997 fieldwork, the society had a branch in Borovo, a village with about 100 residents and one telephone. This small village is perched high on a sunny hillside, along a one-lane road that leads to a popular religious site several kilometers farther ahead, at the top of a ridge. The branch leaders in this community are fairly typical rural residents, and they reflect Borovo's demographic profile—at or close to retirement with grown children. On my first visit during the month of August, I had trouble finding two of the three branch officers because they were busy outside the village in agricultural activities—cutting hay or harvesting potatoes. This branch owes its existence to the society's interest in bringing tourist groups to this picturesque village, and all three members of the branch leadership participate in hosting the guests. The society's president approached Borovo's mayor about this activity in about 1995. Additional people in the village were contacted through him, and the branch was eventually formed. One person, for example, was invited to participate because of his potential to contribute to tourism as a beekeeper and as someone knowledgeable about local history. All three members of the branch leadership, two men and a woman, spoke enthusiastically about the society and its tourism activities in their community. Although the village has only one telephone, in the mayor's office, the branch chair said that communication was not a particular problem because the society's president had frequently visited on his travels between Chepelare and Sofia. They are unsure, however, about what will happen to the branch and its activities in the future, since the society's president has died and he was the person with whom they had had the most frequent contact. Besides showing the enthusiasm of this branch's leadership for the BSCRM and the fact that the leaders are more or less typical village residents, this example reiterates the important role of a single, key individual in maintaining the connections between the local branch and the larger society.

The situation with the BSCRM branch in Shiroka Luka is somewhat different. Shiroka Luka is a village of about 900 residents along a road between the towns of Smolyan and Devin. It is home to a school for traditional folk music and an orphanage. Several houses in the village have been restored as examples of typical regional architecture. Besides participating in the society's tourism project mentioned above, this branch has activities of its own, such as installing signs with information about the local environment and supplying trash containers to encourage visitors not to litter. The branch in this village was established in the early 1990s, and its chair is a childhood friend of the society's founding president, who asked him if he was interested in being involved with the organization. This is an example of the important role played by social relations—in this case connections between *blizki* or close people—in the work of Bulgarian NGOs. The society's president explained to me that the best way to do business was to work through such personal networks. As in Borovo, my initial attempts to contact the branch chair were unsuccessful because on the day I called he was outside the village harvesting hay. This points out that although the chair has a public-sector job and runs a small coffee shop, he also is directly involved in the agricultural economy. The other branch officers were described as being a couple of foresters, a biology teacher, an agronomist, and a retired "nature lover." Most of them were reportedly hunting on the day that I eventually managed to arrange a visit and consequently were unavailable to meet with me. Thus, most of the branch's officers have an occupational association with environmental or natural resource issues beyond their involvement in smallholder agriculture in the postsocialist period. This also makes the point, also raised by Humphrey (1998:viii) and in the Zaburdo case study, that many people living in postsocialist villages are not just peasants or smallholders. They frequently have professional or vocational training and did not necessarily identify themselves as farmers before the political changes made private farming a matter of survival. The branch has additional members, some of them students from local schools. It meets several times a year, and the meetings are often arranged spontaneously on the street or in the chair's coffee shop. This example illustrates how this local branch is embedded in the community and how preexisting social ties played a role in its establishment. In contrast to the Borovo example, members of the branch's leadership in Shiroka Luka have a more direct professional association with the environment, reflecting in part the larger size of the settlement and the associated nonagricultural employment opportunities. This branch also appears to have been less affected by the death of the society's president, perhaps due to the community's larger population base, several more years of experience as an organization, and involvement in locally initiated and controlled activities.

Making Connections and Working with Others

The NGO literature is of two minds concerning the relationship between outside NGOs and the rural communities in which they sometimes work. The optimistic version of the argument is, first, that intermediary NGOs or grassroots support organizations— national or regional groups operating within a domestic context— play valuable roles in supporting or energizing local groups and in providing them with links or connections to other political and financial powers (Carroll 1992; Fisher 1996); and, second, that international environmental NGOs themselves play important roles in linking biological and physical environmental conditions to the political realm at both the local and global levels (Princen and Finger 1994; see also Murphree 1994). Others are less optimistic about the role of national and regional NGOs in promoting such connections. For example, Arellano-López and Petras (1994) express concern that reliance on international donors for financial support will weaken alliances between workers and peasants, on the one hand, and the urban and professional classes who staff and direct most of the NGOs, on the other. They also suggest that in their role as brokers between local-level or grassroots organizations and the national government or international organizations, intermediary NGOs increase the political and economic isolation of local communities.

This section examines the roles played by Bulgarian non-governmental environmental organizations in linking global concerns about biodiversity conservation with local realities—that is, with local-level natural resource use and management in the areas designated for protection and with the lives of people residing in these settings. In particular, this section analyzes connections between regional or national-level NGOs and international organizations, and also between regional or national-level organizations and rural communities or community groups.[7] As Fisher (1997) points out, such connections can involve flows of information, people, funding, and ideas.

East–West Links in Bulgarian Biodiversity Conservation

International conservation organizations and foreign national NGOs with international programs play a role in the biodiversity conservation efforts of Bulgarian NGOs. These links, which are typically East–West, can involve the provision of financial support, technical support, or both to Bulgarian organizations, as well as more informal contacts for information exchange and support of various initiatives. Except for the participation of a few individual scientists on IUCN commissions, the substantive involvement of these international or foreign organizations in Bulgaria postdates the 1989 political changes. The examples in this section focus primarily on links in terms of the funding, personnel, information, and expertise that are exchanged between some of the more prominent foreign organizations and Bulgarian groups, along with some specific impacts of foreign projects at the local level. Other foreign-funded projects more oriented to local communities are discussed later in this chapter.

In the early 1990s, two Bulgarian NGOs received funding from the World Wide Fund for Nature (WWF) for conservation projects in the central and western parts of the Rhodope Mountains.[8] A WWF official explained the organization's general assistance strategy to me as follows: To the extent possible, WWF works through and with existing organizations in a given country. In its relationships with domestic environmental groups, it provides financial support and also tries to enhance the competence of local or national NGOs through training and technical assistance. The WWF's capacity-building efforts for NGOs, as occurred in Eastern Europe in the early 1990s, included three types of financial assistance. It began with grant programs for individuals to improve their skills, for example, through participation in short courses. Once contact was established with an NGO through such individual skill-enhancement programs, WWF typically started getting requests from the group for other kinds of assistance, such as setting up an office with computer, fax, copier, and so on. Consequently, the second type of assistance was equipment grants, usually to small organizations they had worked with previously. The third type of assistance was the provision of project grants once WWF believed that the NGO had the capacity to handle a project. In the early 1990s, WWF extended a standing invitation for organizations to propose projects, and such was the case with two projects described below. Funds have since become more scarce, however. Consequently, the international organization has a more defined program of its own and only funds projects that are part of this program. The Rhodope Mountains fall geographically within the organization's initiative regarding 200 priority ecosystems, although no specific projects were underway in the Rhodope through this initiative during my fieldwork.

During the early 1990s, however, the Wilderness Fund received support from WWF for a project in the western Rhodope. My concern here is the financial relationship between the two organizations, and additional details of project activities are discussed later. According to a Wilderness Fund official, the organization's dependence on WWF financial support created problems from the very beginning because the long-term nature of the project was not initially clear. Instead, the organization was signing contracts with WWF on a yearly basis and consequently planning was done on an annual rather than a long-term basis. Along with organizational issues, the planning issue was seen as problematic for the Wilderness Fund's relationships with the rural communities in which it worked. Wilderness Fund representatives would establish useful relationships, start a project, and then when the funding ran out in a year or two would lose contact with the local people. In the late 1990s, I discussed with Wilderness Fund officials their connections with WWF. By this time they were unclear about the status of their relationship with the international organization, perhaps in part due to personnel changes on both sides. The above-described changes in the way WWF funds projects may have further complicated this relationship if the modifications had not been clearly communicated to Wilderness Fund officials.

The Bulgarian Society for the Conservation of the Rhodope Mountains also received support from WWF, in this case for office equipment as well as for specific projects. An article in the organization's 1994 newsletter spoke enthusiastically of this relationship. Several members of the society had visited Switzerland, WWF staff had visited the Rhodope, and a participatory rural appraisal was conducted in a small Rhodope village with WWF financial and technical support (Bulgarian Society for the Conservation of the Rhodope Mountains and World Wide Fund for Nature 1995). This last activity provided training to the participants and collected information on such topics as resource use and community structure that are potentially useful for planning conservation activities in the Rhodope. It is not clear what use has been made of this information and training. Problems were encountered with involving local villagers in the exercise, and one NGO participant I spoke with questioned the relevance of such an activity for the Bulgarian context. By the time of my fieldwork, BSCRM's connections with WWF had decreased, perhaps again related to the personnel changes and retrenchment of funding procedures mentioned above. The society is not currently involved in any WWF-funded projects, although its newsletter continues to acknowledge the international organization's support for newsletter production, including the provision of computer equipment. (I myself had trouble figuring out how project-management responsibilities were distributed among WWF's various offices, and when I visited WWF's international headquarters in Gland, Switzerland in 1998, finding someone to tell me about these projects of not so long ago proved difficult.)

Three other environmental organizations based in Switzerland are involved in biodiversity projects in Bulgaria through their participation in the Bulgarian-Swiss Biodiversity Conservation Program. The primary contractor is the Swiss League for Nature Protection (ProNatura). The World Conservation Union (IUCN) and the Swiss Association for the Protection of Birds (SVS) are also involved in program implementation. Representatives of the Swiss NGOs make frequent trips to Bulgaria—one of them told me that he visited perhaps four to six times a year. These frequent contacts may be facilitated by the relative geographic proximity of the two countries and the resources involved in this multiyear, government-sponsored program. In addition, program oversight is provided by an individual attached to the Swiss embassy in Sofia. Several projects under this program have components relating to the human communities in and around the project areas through the promotion of sustainable development and information provision. Bulgarian NGOs manage three projects under the Bulgarian-Swiss program, and participation in this program means a fairly steady, multiyear flow of funds for the Bulgarian organizations. On other projects, the Swiss NGOs work with Bulgarian government agencies, although here, too, the intention is to incorporate local organizations to the extent possible. The projects involving Bulgarian NGOs are described below to illustrate the relationships involved and problems encountered along the way.

The Swiss Association for the Protection of Birds works with the Bulgarian Society for the Protection of Birds (BSPB) on two of these projects. Although SVS is primarily a domestic, membership-based organization, it has a tradition of international projects because of the migratory behavior of birds and their consequent movement across different countries without respect for borders. (One of the two most important bird migration routes in Europe, the Via Pontica, crosses Bulgaria [Michev and Iankov 1998:417].) The Bulgarian Society for the Protection of Birds has been involved in this program from the beginning in proposing projects in the Eastern Rhodope and at Poda Lagoon, near Burgas, for inclusion in the Bulgarian-Swiss program. (On other wetland conservation projects along the Black Sea Coast, SVS works with the Bulgarian government.) The Poda Lagoon project focuses on more traditional conservation and protected-area management activities, but it also includes the establishment of an information and education center. Poda Lagoon is the first protected area in Bulgaria to be proposed by and then managed by an NGO. For its work on this project, BSPB was the first Bulgarian organization to receive the Henry Ford European Conservation Award, which came with a $5,000 cash prize. Along with its more scientific conservation and data-gathering activities, the Eastern Rhodope project is concerned with promoting sustainable development as a way to ensure safe nesting sites and non-contaminated food supplies for the birds of prey for which the area is famous. An information center has been set up in the town of Mandjarovo for both the local population and foreign visitors, giving the project a local presence. The SVS conservation officer for the Bulgarian-Swiss program feels that working with Bulgarian NGOs was a good choice. He said that the NGOs are weaker than had been anticipated, but that they were still more effective than the Ministry of the Environment.

The third project in the Bulgarian-Swiss program that is managed by a Bulgarian NGO concerns highland pastures of the Central Balkans National Park. Here the Swiss League for Nature Protection works in cooperation with the Wilderness Fund, which, in turn, has involved organizations in communities around the park in some activities. Along with biologically oriented data collection, community-oriented components of the project include assessing pasture use by local communities, establishing contacts with local authorities, implementing information campaigns about the existence of the park, and evaluating the potential for reintroducing a local sheep variety as a way of maintaining the pastures. Unlike the two previous examples, the Bulgarian NGO was not formally a partner from the beginning of the project. There was an informal partnership, however, because a few members of the project management team also worked with the Wilderness Fund, and the organization was subsequently formally recognized as the NGO partner. (It is not unusual to find individuals who are employees or members of NGOs working in government-sponsored conservation projects.)

Besides its work on the Bulgarian-Swiss program, the Bulgarian Society for the Protection of Birds is the BirdLife-International partner organization for Bulgaria and as such is paired with the United Kingdom–based Royal Society for the Protection of Birds (RSPB).[9] Since 1992, the Royal Society for the Protection of Birds has provided financial support for BSPB's main office, paying for a small staff plus office accommodations. These funds are provided in exchange for work by the Bulgarian organization on the

BirdLife program—data collection regarding conservation of species and sites. This funding is accompanied by other forms of support for the organization, with the country director for Bulgaria visiting BSPB a couple of times a year, plus visits from other specialists as necessary. Conversations with people in both organizations suggest that these relations are cordial and ongoing—perhaps related to the ongoing nature of the BirdLife work program. The Bulgarian organization can also apply to RSPB for special project funding. The goal in providing this support is for BSPB to eventually become independent; however, this may take several more years in the Bulgarian case. According to BSPB officials, they have also benefited from an exchange of ideas with their partner organization about such things as strategic planning—a concept that was not popularly known in Bulgaria and which they have found beneficial. Another aspect of this partnership is that once RSPB establishes a relationship with a particular organization in a given country as the BirdLife partner, they tend always to work through that one organization. If other organizations in a country write to RSPB, they are referred to the already-established partner, although the referral is not always popular with the other organizations.

Along with direct funding from Western conservation NGOs for institutional development or projects, Bulgarian environmental organizations benefit in other ways through their associations with other international groups. The Wilderness Fund, for example, is the only NGO member of IUCN in Bulgaria and as such has participated in meetings of the union.[10] (Bulgaria's Ministry of the Environment and Waters is also an IUCN member, and BSPB has informal contacts with the international group). Although this affiliation costs the Wilderness Fund money, the connection is important for contacts, information, and resources. One example cited in this regard was information about protected-areas legislation in other countries. In writing comments on the draft of a new protected-areas law for Bulgaria, for example, Wilderness Fund members consulted IUCN's Commission on Parks and Protected Areas and its Environmental Law Commission about how the proposed Bulgarian law compared with other European laws.

These examples illustrate some of the connections between Bulgarian NGOs and international or foreign-donor organizations in the promotion of biodiversity conservation in Bulgaria.

These connections are critically important for providing funding for the activities of Bulgarian NGOs in the postsocialist period. Yet, the examples of funding from WWF also illustrate that such funding is not necessarily unproblematic when its short-term or project nature affects the way projects are managed or otherwise influences NGO relations with rural communities. These East–West connections can also result in the exchange of information and ideas and in visits of organization personnel to the respective countries. It is perhaps noteworthy that all the Bulgarian organizations discussed in this section have been around since at least 1990, have a presence in Sofia, and are staffed or otherwise coordinated by multilingual environmental professionals. Although the most obvious flow of information and funding is from west to east, these connections can benefit the Western organizations as well. One example is the contribution of the Bulgarian Society for the Protection of Birds to the BirdLife data-collection program, and another is the experience gained by the Swiss Association for the Protection of Birds in working on its largest international project—the one in Bulgaria—where it can accomplish a substantial amount of work for considerably less money than it could in Switzerland.

Bulgarian NGOs Working at the Local Level

Finally, what about the relationships between the national or regional groups in Bulgaria, on the one hand, and local groups and local communities, on the other? How are Bulgarian groups addressing issues of concern to rural Rhodope communities in their biodiversity conservation activities? The first point to make is that such project components are not uncommon. Bulgarian environmental organizations and activists often pay attention to issues affecting local residents in their conservation activities in terms of such things as environmental education and promoting environmentally friendly, income-generating activities. One example of efforts to address local-level human–environment relations comes from a project to protect the Bulgarian wolf population. In explaining the project, an NGO representative said that one goal was to develop ways to help livestock owners protect their herds and flocks from the wolves that still roam parts of the Bulgarian countryside. One such idea was to reintroduce a local dog variety that is particularly effective in protecting the livestock. The dogs would be bred, and then some of them would be sold to urban dwellers as watchdogs—meeting a demand due to what is perceived as an out-of-control crime rate. Others would be given to herders to help protect the herds and flocks.[11] Similarly, the multicountry, multi-NGO project for the conservation of the bear population on the Balkan Peninsula includes an investigation of attitudes toward bears in hopes of understanding why bears have survived in the Balkans but not in other parts of Europe (i.e., is it just geographical or is there something in the attitudes or practices of the local residents). Other project components include a public awareness and environmental education campaign in areas where bears are found and the collection of data on the bears themselves. Three kinds of activities that conservation NGOs use in attempting to influence or otherwise relate to rural communities are environmental education and public awareness activities, the promotion of medicinal-plant cultivation, and ecotourism.

PROMOTING ENVIRONMENTAL AWARENESS AND
ENVIRONMENTAL EDUCATION

Environmental education and the promotion of environmental awareness are popular activities for many Bulgarian NGOs, and such efforts are often incorporated into conservation projects. Frequently, although not always, this means environmental education programs for schoolchildren and university students. Time and time again, NGO leaders told me that they focus on young people, because children are seen as the future—or conversely, the future is seen as belonging to the children. Younger generations are also thought to be more open to new ideas and more likely to change compared to their parents, who grew up under the socialist system. In some cases, NGOs hope that they will be able to influence the parents through their children. Representatives of several organizations, for example, told me of children's drawing contests in which the parents knew the themes and became interested in the projects through their children's participation. Adults may be targeted as well, such as through posters that one NGO produced encouraging mushroom collectors to harvest the fungi in a more sustainable fashion. Now that many Rhodope forests have been privatized, a couple of NGOs are interested in promoting sustainable forest management.

One example of such an environmental education effort comes from the public awareness component of the Wilderness Fund's project on conservation in the western Rhodope in the early 1990s. Public meetings, slide shows, and meetings with farmers held in the town of Batak and surrounding villages were not well attended. There were, however, many entrants for a children's drawing contest in 1994 on the theme of nature in the Rhodope. The winning children received prizes, and some of the winning pictures were used to produce a calendar. In addition, the organization arranged for some of the pictures to be used as background for the nightly weather report on the television news. The latter is said to have particularly attracted the attention of the parents, who had not necessarily been interested in the contest beforehand. During fall 1997, I met several students who had participated in this contest. They remembered the contest, the prizes they had won, and the pictures they had drawn, but had little sense of the project sponsors. Thus, even though this drawing project experienced better participation than the programs aimed at adults, it is unclear that these efforts had lasting impacts on the local level. One might ask, however, whether the impact would have been greater if NGOs had been better known entities at the time or had a more lasting presence in the community.

PROMOTING THE CULTIVATION OF MEDICINAL PLANTS

Bulgaria's long history of medicinal-plant exploitation has not escaped the gaze of environmental NGOs. The Wilderness Fund's WWF-sponsored conservation project in the western Rhodope involved three components and took place in the Batak region during the first half of the 1990s. The WWF's participation in this project was primarily through its provision of money. Wilderness Fund officials selected Batak for the project's community-oriented activities because it is one of the largest population centers in the project region and also because of the organizational possibilities associated with the municipality. First, organization members periodically passed through the town on the way to a nearby field-research station, and, second, they had acquaintances in Batak who could assist with the project, particularly a biology teacher at the local school who played a key role in the drawing contest described above. She had studied under an NGO member when she was a university student in Sofia. The public awareness and environmental education component of this project was discussed in the previous section. The project's second component involved reviewing existing literature and conducting field research on the area's natural values to identify important areas in the western Rhodope for conservation. They then drafted a proposal to the Ministry of the Environment, suggesting the establishment of one or more new protected areas in the region. Such scientific research is another common activity of Bulgarian environmental NGOs, which is not surprising given the technical expertise of their members. A protected-area proposal was developed, but no action was taken on it or other proposals for parks in the Rhodope during the 1990s because of concern about the impact that the restitutions of agricultural land and forests in the region might have on attempts to create a national park or other protected area.

The project's third component was a pilot program promoting the cultivation of rare and threatened medicinal plants for commercial harvest. The idea behind this project, which came from Wilderness Fund members in consultation with other medicinal-plant experts, was to support local livelihoods by providing an alternative income-producing crop suitable for marginal mountain lands. It was also intended to reduce collection pressure on natural wild populations of relatively rare medicinal plants by providing an alternative to their collection in the wild.[12] Besides working with local farmers on plant cultivation, the project was involved in product marketing, and eventual establishment of a processing facility was envisioned. Some Wilderness Fund members involved in the selection of plant varieties for the project also held professional positions at the Institute of Botany at the Bulgarian Academy of Sciences, which is mentioned because of an association made in the community between the project and the Academy of Sciences. The project coordinator, who lived in Plovdiv, about an hour's drive from Batak, had studied medicinal plants as a university student and was subsequently attempting to break into the medicinal-plant business. Wilderness Fund officials met him through some colleagues and thought he would be good for the job because he could work on both the scientific and business dimensions of the effort.

The pilot cultivation effort took place in a village near Batak that was identified as having good conditions for the project as well as interest on the part of local authorities. Although not necessarily the rarest of plants, Dalmatian sage (*Salvia officinalis ssp. dalmatica*) was selected for this effort because it seemed to offer the possibility of a quicker economic return. Through the village mayor, project officials contacted an interested farmer. They gave the farmer plant seedlings and an advance of money to cover his risk, with the promise of additional money upon a successful harvest. According to the farmer, the plant was easy to take care of and grew very well; it was not adversely affected by the harsh mountain weather.

Perceptions of this project among community members reflect issues raised earlier concerning the lack of visibility of nongovernmental organizations. Most people interviewed about the project, both in Batak and the project village, either identified it as being associated with the Academy of Sciences or thought that it was a private business initiative of the coordinator. In other words, the project was not particularly identified with the NGO funding the project (WWF) or with the NGO implementing the project (Wilderness Fund). Perhaps this is not so surprising given that people in this rural setting lack experience with such organizations, while the Academy of Sciences and private businessmen are more familiar actors. Initially, some people were reportedly suspicious of the project, and rumors circulated about its legality and the source of the plants. Once villagers saw how well the plant was growing, however, interest in it reportedly increased.

Although relatively successful from a biological standpoint, the project was less successful economically, and eventually it ended because of difficulties with selling the resulting herbs. I heard several, not necessarily mutually exclusive, explanations for why the harvested herbs were never collected from the farmer who grew them. The most significant problem concerned locating a buyer. One person suggested that the business groups controlling that particular market niche did not want to let the project coordinator into the market for medicinal plants, and in any case, the need for marketing skills and experience was identified as a problem for the project. Events in the lives of both the farmer and the project coordinator additionally made it difficult for the plants to be picked up. Another difficulty related to the time frame of the project, which involved two separate and relatively short phases. One Wilderness Fund activist involved in the project suggested that five or ten years would have been a more appropriate project duration than the two separate phases, each approximately a year long. He added that it had not been clear at the outset that there would even be a second phase. By the time local residents were interested in participating in the project, he said, it had ended. Yet another problem mentioned by one individual was personnel changes in both the implementing and funding organizations, which created problems for continuation of the project and its funding. He said that the Wilderness Fund had been interested in continuing work on medicinal-plant cultivation as a free-standing project, but was unable to get funding to do so. In the meantime, the Bulgarian NGO has become involved in other projects, especially as part of the Bulgarian-Swiss program, and its energies are heavily directed into that activity. Even so, some Wilderness Fund members are interested in continuing work on medicinal-plant cultivation, perhaps expanding it to other parts of the Rhodope.

These explanations for the project's lack of success suggest several important things regarding community–NGO relations and the conditions for such activities in postsocialist Bulgaria. While the project was well intentioned, with rural residents in mind, it was not successful due to factors external to the village. The project nature of the funding made project planning and establishing a longer-term presence in the community difficult. (Recall here, too, the other examples in these chapters of projects and organizations that have not achieved sustainability due to current conditions in Bulgaria.) Now that the project is over, NGO officials have limited knowledge about the current status of the cultivation effort, and the local participants have lost touch with the NGO officials. In addition, this example illustrates the way in which the NGO sought outside assistance regarding an aspect of the project with which it lacked internal expertise—in this case, marketing—although in the end this was not enough. Finally, the economic situation in Bulgaria and organization of that sector of the economy may have created difficulties for the marketing end of the project.

EVERYBODY IS TALKING ABOUT ECOTOURISM

As with many other places around the world, ecotourism is being looked to as an environmentally friendly economic activity for Bulgaria. Although most international tourism in the country focuses on package tours to the Black Sea Coast and mountain ski resorts, some level of ecotourism predates the political changes of 1989–1990 and formation of most environmental NGOs. Because of the country's rich bird life, bird-watching groups have been coming to Bulgaria for 10 to 15 years (Yankov 1998), and participants often give money for conservation projects to the government or more recently to the Bulgarian Society for the Protection of Birds. Thus, it comes as no surprise that another way in which several Bulgarian NGOs are seeking to link the promotion of local livelihoods with environmental protection in the Rhodope Mountains, as well as elsewhere in the country, is through ecotourism. Some local governments are also involved in promoting this activity, although national tourism officials have been slower to become involved.

In describing the tourism activities of the Bulgarian Society for the Conservation of the Rhodope Mountains, its president explained that people prefer to participate in activities that make it possible for them to survive economically. Since 1994, the society has worked to establish ecotourism activities in several Rhodope communities, including the above-mentioned villages of Borovo and Shiroka Luka. The idea behind this, he explained, is that if local people start to ask where the ecotourism money comes from, the answer will be that it comes from the natural and cultural heritage of the region, and hopefully they will then recognize the value of preserving this heritage. (Representatives of several other Bulgarian NGOs provided approximately the same explanation for how tourism could help both the environment and local people.) As evidence of local support for this activity, he reported that many residents of villages in which the society has tourism activities have joined the organization. In its tourism activities, the society works with a Swiss ecotourism company, receiving in return a portion of the proceeds to use in their projects, thereby getting around the barriers to NGO participation in business activities discussed in chapter 1. Reflecting the organization's concern for local people, a portion of the profits from this activity and donations from the tourists are given to the participating communities. New tires were purchased for the ambulance in one village, for example, and funds were given to buy furniture for another village's cultural center. I spoke with several people who hosted tourists through the society. They were enthusiastic about this activity and the organization and had enjoyed the contacts with their foreign guests. Although they appreciated receiving funds from the larger organization, saying that community members had joined the local branch to show their support, these leaders were not always aware that the funds were specifically from the tourism activities, instead understanding them to be from the society generally. Their level of participation also clearly depended on the guests coming to them through the tourism company or the president's other contacts, since the village branches were not necessarily set up to make the arrangements for such activities themselves.

Another organization involved in the promotion of ecotourism in the Rhodope is the Bulgarian Association for Rural and Ecological Tourism. This small organization initially began its work under the auspices of the Bulgarian Geographical Association before completion of the paperwork necessary for its status as an independent organization. Its office is in the building of the Institute of Geography at the Bulgarian Academy of Sciences, where its president works. Among the places where this group has focused its efforts is an area north of the Balkan Mountains, where its president has a family home, and Trigrad, a south central Rhodope village where the geographers often conduct research projects. Its activities include holding seminars, talking to local people about the idea of rural and ecological tourism, constructing ecotrails, and holding press conferences to promote its activities and the idea of rural and ecological tourism generally. It recently received funding from the Democracy Network Program to hold a series of seminars on village and ecological tourism in several parts of the country, including the Rhodope.

I twice accompanied organization activists on their travels in rural areas and have spoken with rural officials contacted by the association on other occasions. The organization's representatives usually seem well received by the local officials they contact. A common, although not the only, strategy they employ when going to a new community is to contact its mayor. Indeed, one Rhodope mayor has attended events the group has sponsored in Sofia and northern Bulgaria, and he has also sought its input on a project proposal he was developing. On several visits to the association's offices I have either met local government officials or was present when phone calls from such officials have been received. Officials in another Rhodope village I visited were interested in organizing their ecotourism activities in association with the NGO, rather than through a Bulgarian adventure travel agency that has sent visitors to the village, because they believed that greater profit would come from avoiding an intermediary agent. Yet, there is a dilemma here in terms of what the NGO can do by itself in tourism promotion. In the case of the village just mentioned, no tourists had come through its association with the NGO. The association comes up with ideas, works on building trails, and tries to introduce local officials and residents to the idea of village or ecological tourism. Yet, it is not necessarily able to organize—nor is it interested in organizing—large trips to these communities.

More recently, Green Balkans has begun work on ecotourism promotion in the central Rhodope, with funding from the European Union through the PHARE partnership program. This activity is part of a joint project involving four environmental NGOs in three Eastern European countries—Romania, Bulgaria, and Hungary. The Bulgarian project includes ecological research into the natural values and biodiversity of the project area near the Arda River; promotion of alternative, sustainable, and hopefully more lucrative agricultural activities (i.e., berries, local livestock breeds, beekeeping); and support for ecotourism in the region. In the latter case, Green Balkans works with eight Rhodope communities. Early in the project, local authorities from these villages traveled to Hungary with Green Balkans activists for a workshop and meetings with people from the other project areas. At follow-up community meetings in the Rhodope villages, Green Balkans officials suggested creating community-level NGOs with 10 to 15 members each to work on tourism promotion or other environmental issues; they also suggested cooperation among the several village-level groups in the establishment of a joint tourist bureau to save on marketing expenses. Rather than creating local branches of the Green Balkans, the organization's intended role is acting as a catalyst by suggesting the creation of the groups and then sharing experience on topics such as NGO registration, fundraising, and project management.

A community meeting that I observed one snowy winter afternoon seemed well attended given the village size, and the people attending appeared interested in the ideas presented. Indeed, one young man with a new house was frustrated to hear that it would be several months before project officials could come back to evaluate the suitability of his house for hosting guests. The Smolyan-based project coordinator reported that they were pleased with the response to these first meetings, which took place in several villages, and he moreover noted that the response had been better in those communities that had had previous experience with tourism. On a more recent trip to Bulgaria, I learned from Green Balkans that local societies have been established in four villages included in this project and that they were applying for funding to implement a similar program in additional municipalities. Activists in one of these municipalities told me about a meeting held on the subject and said that the community seemed generally favorable to the project.

These examples show the range of ecotourism activities in which Bulgarian NGOs are involved in the Rhodope and several different ways that such activities are organized. The latter include working with a commercial tour operator, encouraging the establishment of community-based organizations to work on tourism promotion, and promoting the activity without necessarily suggesting an organizing structure. My impression from talking to local residents who had been contacted by these organizations was that they were open to participating in and thereby gaining income from such activities. Some Rhodope residents are aware of international interest in the Rhodope region through their occasional contacts with foreign tourists, whether as part of NGO-sponsored or commercially organized tours or as independent travelers both during and after the socialist era. Several times during my stay in Zaburdo, for example, tour buses full of foreign visitors stopped at a village coffee bar after taking the tour groups to see the nearby Wonderful Bridges. In other years, after visiting these natural stone arches, NGO-organized tour groups stopped at the house of a local weaver to see her handicrafts and watch her weave.

Several issues merit consideration, however, in discussing the feasibility of ecotourism in Bulgaria and the involvement of environmental organizations therein. Tourism infrastructure is lacking in many rural areas, and consequently foreign tourists not heading for large, established resorts may find it easiest to travel with a translator and arranged transportation—thus influencing the shape of this activity at least in the short run. In addition, rural tourism is not necessarily going to be the economic savior for all villages. First, there may not be sufficient demand to send many visitors to numerous villages; second, careful examination of tourism's environmental impacts would be necessary if there were such high demand; and, third, money and time may be needed for infrastructural development. As well, this activity is highly seasonal—predominantly during the summer, and in a few areas during the winter ski season as well—and there is a potential for conflict between the peak agricultural season and the peak tourist season. Time would need to be devoted to attending to the tourists, and yet at the same time, if the hay was not harvested, there might not be any milk to give the visitors to drink. Another important issue is exactly what the NGOs do regarding tourism—that is, their specific role in this enterprise. Do they become involved in the actual implementation of tourism, as may now be possible given the evolving interpretation of their rights to engage in business activities? Or do they simply promote the activity, perhaps constructing or maintaining trails or producing brochures describing local landmarks, and possibly provide technical assistance as in the Green Balkans project described above? These questions cannot be resolved here, but certainly merit consideration in evaluating NGO involvement in ecotourism and its connection to conservation efforts.

The preceding examples of environmental education, medicinal-plant cultivation, and ecotourism illustrate some of the ways in which Bulgarian environmental NGOs attempt to work with rural communities in the Rhodope to achieve ends in relation to biodiversity conservation. So too do examples mentioned in passing concerning promoting the raising of local animal breeds and higher-value alternatives to the potato. There is global interest in the Rhodope and in community-based conservation, and the NGO activities described above demonstrate in principle NGO concern for rural Bulgarian communities. Yet, NGO presence in the region is not strong, and the results of projects on the ground were mixed. One question is the extent to which the organizations, their activities, or both have an ongoing presence in the rural areas. Another consideration is the way in which local people view the urban-based groups. It is relatively rare to find national or regional groups working in ongoing partnerships with local groups; to the extent that such cooperation occurs, it tends to be more common on a project basis. (Another example of this cooperation is seen in the next chapter.) There are a few examples of such partnership, however, such as the Green Balkans' ecotourism project, and some of the larger organizations do have local branches in the Rhodope. Of course, working with local groups either requires the existence or establishment of local groups, which is the route taken by Green Balkans in their work in the southern Rhodope. The urban-based groups do not necessarily have a strong presence in rural communities. This was illustrated in the case of the Wilderness Fund's western Rhodope project, which was more commonly identified with the Academy of Sciences or as a private business venture. (Part of the problem there was a lack of awareness of what NGOs are.)

Finally, when rural residents are aware of the groups, local people do not necessarily recognize NGOs as the "best people" for the job when it comes to resource management. This returns to the issue of legitimacy raised earlier. The topic of conservation organizations came up unexpectedly in a dinner-table discussion during my visit to a hunting reserve in the western Rhodope. After a London hunter's successful day—he killed a chamois before lunch and a wild boar after lunch—we sat down for dinner in the lodge's formal dining room. Sitting at the table were the hunter and his translator, me and my research assistant, and the hunting guides, two brothers who had lived their entire lives on the reserve. As the conversation turned to sustainable resource management, the hunter spoke of his observations about the effectiveness of the Communal Areas Management Program for Indigenous Resources (CAMPFIRE) in southern Africa, where he had hunted elephant, and subsequently the topic of NGO involvement in wildlife conservation came up. The hunting guides, who also managed the hunting reserve, questioned the value of giving funds to organizations in Sofia, naming a couple of them, and suggested that the money would be better given to people like them who had hands-on experience in wildlife management. Thus, even if local residents are aware of NGOs, some of them may feel that city people should not tell them how to do their jobs. If nothing else, this suggests that some distance still exists between urban NGOs and local resource users.

Linking the Global and the Local

This chapter has explored the ways in which environmental NGOs are involved in linking the global and the local realms in biodiversity conservation efforts in Bulgaria and especially in the Rhodope. Given the small size of many organizations, it is not surprising that a few individuals often play key roles in keeping NGOs or their local branches going and that when such individuals move or their attention is otherwise diverted, an NGO's activities can decline or even cease. Perhaps not so expected is the importance of the identity of some organization leaders from a civic or professional standpoint in giving legitimacy to organizations in the postsocialist Bulgarian context in which NGOs are not always understood to be legitimate and effective players in environmental policy and practice. Although it is more common to find Bulgarian environmental NGOs in cities and larger towns than in the countryside, such groups do exist in these rural settings, and their members and leaders are to a large extent local people. An examination of some of the individuals involved in NGOs in small Rhodope communities and the circumstances surrounding the establishment of these rural organizations shows the diversity of the men and women involved, ranging from rural farmers to professional foresters to business entrepreneurs. It also demonstrates the importance of informal social networks in the establishment of the rural branches or organizations as well as the influence of external opportunities, such as participation on international study tours.

Location does have significance for Bulgarian environmental organizations, the majority of which are located in urban centers. Being situated in the nation's capital can be an advantage to organizations in terms of the access it affords to foreign donors as well as to Bulgarian decision makers and decision making on the national level. At the same time, however, an urban location can mean that a conservation organization is removed from the rural areas it seeks to protect and the people who live there. (Some members of city groups do have close ties to or contacts in rural communities, however, which can both facilitate their work in these settings and make them more sensitive to the local context.) In contrast to the city-based conservation organizations, which tend to focus on issues of biodiversity, rural groups more often include the issue of economic sustainability in their concerns. Indeed, some rural organizations that focus on sustainable development may scarcely resemble the city-based environmental NGOs. They are working to protect the environment as they see fit, based on their local environmental knowledge combined with their concerns about local livelihoods and community sustainability.

Stepping back for a moment to take in the larger picture, Bulgarian NGOs do play some role in linking global concerns about biodiversity with conservation activities at the local level. Several examples have been discussed in this work regarding the participation of international organizations through the provision of technical and financial support to Bulgarian organizations. Given the financial situation in the country, such external support has been important for conservation efforts by both the government and the NGOs. Indeed, one of the interesting points to emerge from these descriptions is that while the NGO community in Bulgaria is not necessarily strong or self-sufficient, the government's conservation infrastructure is weak as well, and thus NGOs have played a particularly strong role in the Bulgarian-Swiss program. Some of the ways in which the Bulgarian organizations have made an impact at the local level are management of the Poda Lagoon protected area, the construction of nesting sites for birds, and feeding the birds. While NGO activities also show a concern for community interests in the Rhodope landscape, the community-relations efforts described did not necessarily have a lasting impact, and a project to promote medicinal-plant cultivation was not successful from an economic standpoint. Ecotourism activities are more difficult to evaluate. Local communities are clearly interested in this potential source of income given the dire economic circumstances in which many of them find themselves, yet exactly what the NGOs can do in the promotion of this activity is unclear. And then there is the question of how rural residents will put food on the table while they wait for the ecotourism industry to reach them. Finally, Bulgarian NGOs recognize the importance of helping local communities develop sustainable livelihoods, both economically and environmentally, through such things as raising local animal varieties and promoting higher-value crops. Yet, some of these activities are difficult to measure, and my fieldwork did not thus far show a significant impact of NGO activities on local-level resource use and management, at least in terms of the kinds of activities that were discussed for Zaburdo. The next chapter looks at another Rhodope village for a contrasting case in which there is a stronger NGO presence.

6

A Civil Balkan Village?
Cavers and Collective Action

Previous chapters have shown that in Zaburdo, as in many rural Bulgarian communities, the formal organizations of civil society are absent. But some communities appear to fulfill this fashionable Western ideal. The story of some environmental activists in another rural mountain village and the various organizations in which they are involved serves as a novel window for looking at nongovernmental organizations and civil society in Bulgaria. This is the story of a socialist-era caving club that has adapted to postsocialist conditions and taken on new organizational forms; it is a story of local social organization, of external connections, and of the larger legal and economic contexts for NGO activities in Bulgaria. Given the amount of attention—and, perhaps more importantly, financial resources—being directed toward environmental organizations and other NGOs in postsocialist countries, because of their presumed association with civil society, it is important to examine critically the organizations that are the subject of all the rhetoric. Moreover, it is particularly valuable to look at such organizations in rural settings where they are often ignored or nonexistent.

It is not my intention to delve here into theoretical discussions of civil society, a topic about which many articles and books have been written.[1] Yet I should say a few words about the rhetoric concerning this topic in the postsocialist context. As one observer recently wrote, "Though the term can elicit groans when broached at academic conferences, it is still favoured in some influential development circles" (Benthall 2000:1). Many accounts suggest that there were no NGOs and no civil society—in the sense of social space that exists between the institutions of the family, the market, and the state—in the countries of Central and Eastern Europe prior to the 1980s and that such organizations only emerged as socialist regimes there began to collapse. Subsequently, NGOs in the postsocialist period have received considerable attention as one of the presumed foundations of a newly developing civil society. Civil society, in turn, is viewed as an important building block for postsocialist democracy. In some cases, environmental organizations in particular are cited as being especially significant in this triumvirate of NGOs, civil society, and democracy (e.g., Snavely and Desai 1995). One sees this connection being made in the literature of organizations as diverse as the Regional Environmental Center for Central and Eastern Europe, the Soros Foundation, and the U.S. Agency for International Development. My interest here thus pertains to this rhetoric about civil society and its relationship to NGOs versus the empirical reality of postsocialist life, not the multitude of theoretical discussions. The case study of organizational life in one postsocialist Bulgarian village provides the empirical basis for this analysis.

In telling this story of cavers and collective action, I begin with some background on the village in question, aspects of which are significant to the establishment of the NGO. Next I present a social biography of organizational life in the village. This case study then allows for a consideration of what makes this "civil" village different from the other "uncivil" communities that I visited—including the rather similar village (Zaburdo) that was the focus of this book's central case study. It also illustrates the kind of links that can develop between rural groups and national or international agents. Ultimately, my concern is what factors contributed to the development of organizational life in a particular village under conditions in which formal, village-based organizations are generally rare.

Ethnographic Setting: Another Rhodope Village

I have visited the village of Trigrad several times during my Bulgarian fieldwork, most recently in September 2001. Two of these visits were on my own, but on two other occasions I ended up there while accompanying national-level Bulgarian environmental NGOs—part of the story that I want to tell. I also occasionally read about the people and organizations discussed here in Bulgarian newspapers via the Internet. Given the small size of the village, the attention it receives from the press makes it somewhat unusual.

Trigrad is a mountain village with about 900 residents that is somewhat isolated due to its location very near Bulgaria's southern border with Greece and behind a narrow gorge of the same name (see Figure 2).[2] As with other communities in the southern half of the Rhodope Mountains, many of the village's economic ties in earlier centuries were oriented toward the south—in other words, to territory that is now part of the Greek nation–state— and its economic fortunes declined when an international border was more firmly established in the early 20th century, largely cutting off access to the southern markets and winter pastures along the Aegean. Like Zaburdo and most villages in the region, the local economy has historically been based primarily on animal husbandry and secondarily on agriculture. These activities were organized into a cooperative farm during the socialist era. The socialist era also brought small assembly workshops to the village in order to provide its rural residents with year-round, non-agricultural employment. In the last decade, the workshops have closed and the collective farm has been liquidated, returning private agricultural activities to prominence. These are largely organized on a household or family basis on land restituted in the early to mid-1990s, and the resulting produce is important for the household economy.

The village has also become a tourist destination in the last two to three decades because of the narrow and impressive limestone gorge, which is designated as a protected landmark. Birdwatchers visit the gorge in hopes of catching sight of the elusive wall creeper (*Tichodroma muraria*), and a nearby cave, the so-called Devil's Throat, has been developed for visitors with stairs and lighting. The gorge is also located along several popular hiking routes and is near a nationally designated nature reserve and some of the only natural lakes in the Rhodope. During the socialist era, visitors often stayed in a mountain hostel located between the gorge and the village, and more recently several local families have begun to rent out private rooms and offer meals to tourists. (The hostel, meanwhile, is said to have fallen into disrepair.)

One Organization, Two Origin Tales

Moving on to associational life, Trigrad is the home of a caving club, which is part of a network of caving clubs that existed under socialism, along with other state-sanctioned organizations for the promotion of sports and physical activities such as the Bulgarian Alpine Club. The caving club carries the name of a rare endemic plant—silivryak (*Haberlea rhodopensis*)—that often grows on rocks near the entrances to caves in the region. The two stories I have been told about the origin of this organization likely reflect two periods in the club's history. Both are worth repeating for what they suggest about social relations and the identities of the club leaders. In one version, the club was established in the mid-1960s, and its first president was a teacher from the city of Plovdiv. At this time no local residents had been trained as teachers, so teachers for the village school were brought in from elsewhere. Sometimes they were the wives of border guards stationed along the so-called Iron Curtain, that is, Bulgaria's international border with Greece and thus with the West. These outsiders—the teachers and the border guards—were enthusiastic about outdoor activities in a somewhat different way than the village farmers, and they came together in establishing the caving club. But local residents also joined the club, among them men who had done their mandatory military service with an alpine division based in a nearby regional center,[3] and many subsequent presidents of the club were locals. For example, the second president was described as being a Pomak (Bulgarian-Muslim), suggesting that he was a local resident, and the local man who told me this story— one of those with army alpine experience—had been the club's president for a couple of years in the mid-1970s. This university graduate had returned to the village after completing his studies at an agricultural institute in Sofia and then worked on the cooperative farm during the socialist era. Although no longer affiliated with the caving club, this man, now about 60 years old, has been the president of the local hunting society for the last decade and is also the president of a local association for ecotourism. The hunting society is another one of these quasi-governmental, socialist-era organizations (albeit with presocialist roots), membership in which is required of all hunters, while the ecotourism association was registered as an NGO under Bulgarian law in 1997.[4] This shows a continuing pattern of involvement in organizational leadership by this individual, even if the particular organizations have changed. I return to this point about key individuals at the end of the chapter.

According to the second "origin" story, a group of village residents who were born in 1961 founded the caving club in 1983. (People who are born in the same year—*nabori*—constitute a recognized and talked-about category in Bulgarian social organization, particularly at the village level. In etic terms, *nabori* constitute an age-set.) The club was established after a few of these young men returned home following compulsory military service, in which they were in the army's alpine division. One of them, the current chair, is now a local entrepreneur who runs a small hotel. He is well known in regional caving and environmental circles, often referred to simply by a familiar form of his first name. He described himself as being an ecologist for 20 years as well as a caver. To be a caver, one must be an ecologist, he said. Another founding member (also born in the early 1960s) was the village mayor in the early 1990s, representing the Green Party, and he is now a local businessman who owns a coffee bar and is also involved in several other economic ventures. Although these individuals do not appear to be significantly involved in village agriculture, this was not true in every case. A third long-time member, for example, fits the more typical profile of what has been called a peasant–worker (Lockwood 1973). He works in a village store, his wife teaches at the local school, and they also grow potatoes. The latter are consumed by the household and bartered to obtain other goods. This man, who also has an interest in photography and video technology, explained to me that he has known the cave club president since the fourth grade and is related to his wife.

The cave club leadership is relatively young and predominantly male. Besides the thirtysomething individuals just described, for example, the club secretary is in his early to mid-twenties. This outdoor and motorcycle enthusiast, who was working at a local sawmill on the day I spoke to him, had joined the club in the fifth or sixth grade. There were 15 to 20 active members when I visited the village in 1997, ranging in age from 15 to 84 years. Most of them were local residents, the secretary said, although a few were outsiders with villas in the region.

As one might expect, the club activities revolve predominantly around caving, with members taking a leading role in exploring and mapping the numerous local caves and also participating in caving expeditions elsewhere in Bulgaria and other socialist-bloc countries.[5] On my most recent visit to the village, for example, one member recounted his knowledge about caves in Afghanistan gained during a trip there, and another caver had previously mentioned a trip to the Pamirs in Central Asia in about 1990. They also host visitors—often members of other caving clubs—who come to explore the caves in their region. One person explained that the local cave explorations, at least in the beginning, included a hunt for gold and silver believed to have been stashed in caves as people fled the region during times of conflict or other crises. But they also included efforts to discover the secret of the Devil's Throat. "What it takes, it does not return," said one man, explaining that on one occasion 500 cubic meters of timber had been thrown into the river that runs through the cave, but none of it had reappeared where the river resurfaces a few hundred meters outside of the cave.

With Bulgaria's postsocialist economic crisis and associated loss of external funding for such activities, it is now virtually impossible for club members to participate in caving expeditions to distant locations. But they continue to explore caves in the region and to train new members in various caving techniques. On my most recent visit, for example, I was told of a ten-day expedition that had been interrupted by bad weather earlier in the month, and that they were expecting an official from the central caving federation to join them when it resumed. As the local affiliate of the Bulgarian Tourist Union (the umbrella organization for state-sanctioned sporting clubs in Bulgaria), the club also manages the Devil's Throat Cave as a tourist attraction (e.g., collecting admission fees and providing tourists with information about the cave) on behalf of the municipality.[6] Beyond this, it has become further involved in tourism in recent years. Since at least 1997, it has offered guided adventure tours of another nearby cave, involving roped rock climbing and rappelling. If the tourists desire, a member of the club with video skills and equipment can film the trip for an additional fee. The club is also licensed as a tour operator by the Ministry of Tourism, and it recently purchased eight horses so that it can offer riding tourism as well. Several members of the club—7 of 17 members in 2001—work as guides for the tourism activities. The fees charged for the adventure cave tours provide some income for the guides as well as for the cave club—for purchasing needed but expensive equipment, for example. Indeed, its president said that they might be one of the richest caving clubs in the country as a result. Finally, on a couple of occasions, the club has organized concerts in front of the Devil's Throat Cave; the one in 2001 also included an art exhibit inside the cave. Based on the descriptions of its members, this village organization clearly has regular and ongoing activities, even if its members represent a relatively small proportion of the village population.

Connections and Offshoots

But this socialist-era organization is just the beginning of my story. It is also important to consider how this organization is connected with and responding to the larger context for such organizations in Bulgaria and beyond. In the early 1990s, the president of the recently established Bulgarian Society for the Conservation of the Rhodope Mountains traveled to Trigrad to ask the president of the already existing caving club if its members would help his organization. He explained that they would be able to get some money for their cave projects through this association. This sounded like a reasonable idea to the caving club leaders, and a branch of the society was established in the village. This was just one of several local branches of the society, but perhaps the only village-based one to piggyback so directly on an existing organization. The local caving club is a collective member of the society's branch in the village, and the caving club president is the chair of the BSCRM branch. Besides the caving club members, about 15 additional people were reportedly members of the local BSCRM branch in 1997.

These ties appeared to have weakened by 2001, at least partly due to factors beyond the village. When I asked the current president of BSCRM in the summer of 2001 about her contacts with the group in Trigrad, she said that one of the society's activists had recently visited Trigrad to take photos for a brochure. While there he had seen the president of the Trigrad club—mentioning the president by his well-known first name—but this seems to have been the extent of the contact. Similarly, in 2001 I asked a Trigrad resident about NGOs in Trigrad; even though the resident had been identified to me in 1997 as a member of the BSCRM branch in Trigrad, the society did not come up in our conversation.[7]

The caving club has contacts with other organizations as well. Its work with these other groups generally consists of a city-based NGO obtaining funding for a project from an international donor or government source, and then members of the Trigrad club help carry out the project activities that are planned in their region. For example, the Bat Protection Society had a project to determine the number of bats in the country, and the Trigrad cavers assisted by counting the bats in that area. Also mentioned were well-known, national-level conservation NGOs. As I related earlier, two of my four visits to the village were with such national-level environmental NGOs who had gone to the village to talk with NGO leaders and the mayor about establishing eco-tourism and sustainable development activities. Clearly the Trigrad organization and its leader are relatively well known in caving and environmental NGO circles, and this sometimes develops into joint activities. Such projects are somewhat episodic, however. In 2001, for example, the caving club president reported periodic contact with people from other organizations but said that the club was not at the time involved in any joint activities with other organizations.

In a separate development, the final one that will be described here, the village caving club itself was registered as an independent organization under Bulgaria's NGO law (the Law on Persons and the Family) in 1996. This was done at the suggestion of someone from the central caving federation in Sofia, because the club had obtained funding from the British embassy for improving the tourism infrastructure inside Devil's Throat Cave. The money had to flow through a juridical person, which the caving club was not. The club leaders were afraid that if the money went through the municipality, which was such a legal person, most of it would be lost due to a lack of budgeting controls in the local government. Such had reportedly been the case with the money for a previous project; much of it disappeared, thus leaving little funding for project implementation. By becoming a legally recognized, independent organization—in other words, an NGO—the club could receive the money directly, and the assumption was that more of the money would actually get to the cave. Some scholars rather naively analyze the growth in the number of NGOs as evidence of the growth in civil society (e.g., Massam and Earl-Goulet 1997; Snavely and Desai 1995). Yet, this example suggests something rather different about this growth. It shows how conditions in Bulgaria, specifically government financial mechanisms, influenced the structure of the NGO community by promoting the registration of an additional independent organization. In other words, this particular increase in NGO numbers can be traced to the budgetary capacity (or rather the lack of budgetary capacity) of the local government.

There is a second point to be made about this particular development. It concerns the role of chance, individuals, and personal connections in such organizations. In Bulgaria, NGO leaders sometimes comment on the importance of *vruzki* or connections, and such connections clearly played a role in acquiring this funding. The caving club's initial contact with the British embassy was established through a Bulgarian employee there, who sometimes spends his holiday in the village and thereby became aware of the tourism potential and infrastructure needs of the cave. Consequently, the first set of funds came from the embassy itself for tourism support, although later funding came through the British Know-How Fund, which is the UK government's bilateral program of technical assistance for the countries of Central and Eastern Europe. Perhaps the head of the Know-How Fund or the British ambassador eventually would have made it to this village, but the chance connection through this Bulgarian employee certainly facilitated the contact and thus acquisition of the funding.

Explaining Associational Life

Nongovernmental organizations—NGOs—in former socialist-bloc countries have attracted considerable attention in the last decade. Beyond their on-the-ground activities and accomplishments in terms of environmental protection, health care, multicultural education, human rights advocacy, and so on, they are hailed by some as one of the foundations of civil society (and hence, by association, democracy). This interest has been followed by funding for such organizations, sometimes in preference to having government bodies accomplish the same activities. Most Bulgarian NGOs are based in cities and larger towns (see also Sampson 1996 on Albania; Close 1999 on Greece), and they are a relative rarity in rural settings. But there are exceptions, and such exceptions deserve more critical anthropological attention than they have received thus far. Having described here the case of what I call a "civil" Balkan village, I want to recap some of the major points to emerge from that analysis and then to consider what makes this village different from the other "uncivil" villages that I visited. Under what conditions, one could ask, might such organizational life develop?

Not counting the village hunting society, because membership in it is required of all who want to hunt and thus it is not strictly voluntary, there have been as many as three local voluntary organizations with environmental activities in this medium-sized village during the 1990s: the caving club, the ecotourism association, and the local branch of the regional conservation society. In some sense, this may be a case in which there is less going on than meets the eye, with the somewhat passive ecotourism association and the now apparently inactive branch of the conservation society. But compared to your average Rhodope Mountain village, this is still a lot. The discussion below focuses primarily on the caving club, which is the point of departure for the other organizations.

Trigrad's caving club is first and foremost a recreation association, and its primary activities revolve around exploring caves. Some critics might complain about the recreational nature of the caving club. One might ask, what or how does this club contribute to civil society or to environmental protection? Now, I could cite examples here from mountaineering, hiking, and bird-watching clubs in the United States, who work to protect the environment so that their members and the rest of society have a place to go. They build trails, lobby Congress, and alarm the broader public of perceived threats. And hopefully I have already demonstrated that the caving club is sympathetic to environmental causes through its willing to cooperate with Bulgarian environmental NGOs on their projects and also through the cave research.

Plus, there is a more important point to be made. Given the project nature of NGO life in Central and Eastern Europe (à la Sampson 1996), I've concluded that a key to the sustainability of the Trigrad club is the fact that there is an ongoing glue that holds it together. Bulgarian NGOs are often young and not particularly institutionalized. In all settings they need activities and individuals to keep them going, as well as financial resources. In a village context, the kind of ongoing activities needed to sustain the momentum of a strictly environmental NGO are often absent. Recall here the earlier-cited comment of one rural community mayor that they had no environmental NGOs because there were no pollution problems there. But caving in the local area can happen on a regular basis. Cavers spend extended chunks of time together, sometimes literally placing their lives in the hands of their companions. Caving thereby creates a special kind of community; one that is re-created or maintained by regular expeditions, looking at photos, and reminiscing about them.[8] For example, the president of the mid-1970s described the Trigrad club as having an ideal purpose and as uniting people. In short, I think the recreation core of the club is particularly significant to its survival and continued existence in the village setting.

To this foundation, initially built in the socialist era on shared common interests and experiences, other activities have been added periodically in the postsocialist period. Ecotourism is not a surprising development, particularly considering the club's role in managing the Devil's Throat Cave, and more generally, ecotourism is very popular among Bulgarian environmental NGOs as a way of promoting environmentally sustainable development. One might say that the movement into ecotourism is adaptive in that it provides the organization with a source of needed financial resources. Although not a politically oriented NGO (that is, one out to directly affect environmental policy), the cave club also contributes periodically to the larger efforts of the Bulgarian environmental NGO community—for example, by carrying out specific local activities as part of larger environmental projects. In sum, the existence of an organizational base also creates a basis for other possible activities, either alone or in cooperation with other organizations. And these are activities that look more like those receiving the attention of the advocates of civil society.

Even though Trigrad is located on the periphery, some distance from the national capital and other population centers, its caving club is connected to other organizations in at least three different ways. And these connections are important for the club's various activities. First, there are long-standing ties with cave federation colleagues with whom they have gone on expeditions at various points in time or who have come to visit the caves around Trigrad. While some of these individuals are based in the capital city of Sofia,[9] others are based in smaller communities, such as in the central Rhodope town of Chepelare. In the past, funding for club activities came through these cave federation connections. Second, there is periodic collaboration with other Bulgarian environmental NGOs—typically Sofia- or Plovdiv-based ones, with better Western-language skills and fundraising experience, so that they are able to get projects funded. The Trigrad caving club assists these other groups with local aspects of larger projects without formal ties or membership. Contact of this sort is likely facilitated by the fact that the club president is a known figure. And third, the club has connections as a corporate member of a local branch of a larger regional NGO. Piggybacking on an existing club perhaps gives this branch a stronger presence in the village, although the branch seems to have faded from the days in the early 1990s when members of the Trigrad club attended the society's annual meeting. To this list, one could also add the chance connections established through individuals who happen to visit Trigrad as tourists on a whim but end up in conversation with their village hosts. One thing leads to another, and the British embassy is providing funding to build new stairs in the cave.

Funding is also important to organizational continuity. Fundraising can be more difficult for NGOs that are not located in urban centers, but Trigrad's caving club has been inventive and successful. Besides receiving funding through other Bulgarian NGOs for participation in specific projects, it has been able to raise funds through tourism, which it can do as a caving club but not so easily under the old NGO law. Then by registering as an NGO, it can directly receive funds from external donors—like it did from the British embassy. Indeed, this registration was done so that the funding would not be lost in a perceived "black hole" of municipal finance. So here we see an example of a socialist-era club that has been relatively successful in adapting to postsocialist conditions, including financial ones.

Finally, what makes this setting different from the other "uncivil" villages—that is, ones lacking NGOs—that I visited? I have
answered this in part with my analysis about the caving club. But
this still leaves the question of why there was a caving club in that
particular village and not in others. In considering this question,
my comparison focuses on Zaburdo, the remarkably similar village described in earlier chapters. Like many if not most villages
in the region, the economies of both Trigrad and Zaburdo were
historically based on agriculture and pastoralism, and these activities were organized into cooperative farms during the socialist era, with the addition of small assembly workshops. Getting
more specific, both are of a medium size with 600 to 1,000 residents, are historically Bulgarian-Muslim, and are geographically
isolated in the sense that they are some distance from a main road
and even further from their respective municipal centers. Also,
both are located at about 1,200 meters above sea level, which has
implications for agropastoral production. Last, but perhaps most
critically, what distinguishes these two villages from most others
in the region is that each is the closest village to a natural phenomenon that serves as an attraction for both foreign and Bulgarian tourists. In the case of Trigrad, the attraction is the Devil's
Throat Cave; in the case of Zaburdo, the attraction is the Wonderful Bridges, which are three natural stone bridges located about a
two-hour walk (or about 10 kilometers) by road from the village.
These two sites are among the most popular destinations for international tourists who visit the Pamporovo resort, and they also
attract Bulgarian tourists on walking or driving tours.

Three factors are worth considering in comparing the two villages with regard to organizational life. First, throughout my
research on Bulgarian environmental NGOs, the importance of
individuals and especially of individual leaders is quite clear.
These organizations are not institutionalized like big Western
NGOs and often their success or disappearance depends on the
presence, absence, or continued attention of a few key individuals. Although this is the case for many Bulgarian NGOs, it is particularly important in smaller communities. Clearly, a couple of
individuals are important for keeping the caving club going in the
postsocialist period, and to some extent it might simply be chance
that such individuals are located in Trigrad and not in Zaburdo.
With regard to the initial establishment of the caving club, a key

role was by played outside teachers and border guards who were not significantly involved in agropastoralism and thus perhaps were interested in tourism in a somewhat different way than the average villager. Simply put, they might have been more likely to welcome an activity that involved tromping around the outdoors in their spare time than those who worked on the cooperative farm and thus also spent their work hours travelling across and working in this landscape. As teachers they were more educated than the average village resident and thus perhaps more inclined to organize such a club. Recall here, too, that one of the home-grown presidents had a university degree.[10] Relying on outside teachers was not unusual for rural mountain villages in the 1960s, but it was somewhat more unusual to have border guards—this would only apply to settlements rather close to the international borders. In that regard, Trigrad is more unique. In the second origin story, local social organization played a key role in that the founders were age-mates, who had in the meantime done their military service together as age-mates are wont to do, creating perhaps an even tighter bond between them. But this aspect of social organization is common to most villages, and thus could again happen anywhere. Second, alpine experience in the army is a common characteristic of several key members of the Trigrad club, perhaps because such people had gained the necessary technical skills for caving and possibly also an appreciation for such activities during that experience. It is not clear, however, whether this is just another coincidence or if there was some sort of tracking of people from this location into the army's alpine division. The only assignment I recall hearing about from a Zaburdo resident concerned a young man who spent his military service as a shepherd, which was also his occupation on the co-operative farm. And third, it may be that the geological conditions were somewhat better for a caving club in Trigrad when compared to Zaburdo. There may simply be more caves to explore around Trigrad than there are in the Chepelare Municipality, and also more caves with archeological sites and other secrets.

On a final note, religion may also play a role in the formation and sustainability of Trigrad's caving club, albeit in an unexpected way. Like Zaburdo, Trigrad is historically a Bulgarian-Muslim village and somewhat more religious than Zaburdo, at least judging from the fact that Trigrad has a mosque that is being renovated and is used at least occasionally, while Zaburdo has neither mosque nor church. Some scholars question whether or to what extent true civil society can develop in a Muslim society (e.g., Gellner 1994). Some might argue that this is not strictly a Muslim society, since it concerns a Muslim minority in a predominantly Christian nation–state, but I want to make a rather different point. The village's historically Muslim identity might have actually contributed to the formation and continuation of the caving club. First, such communities often had to rely on outsiders as teachers because young people were at times discouraged from leaving these tight-knit agropastoral communities for the education that they would have needed to become teachers (or for other nonagricultural professions). And such nonlocal teachers played a key role in the initial establishment of the caving club. Second, Muslim communities often experienced less rural–urban migration during the socialist era and thus even today tend to have larger and younger populations than comparable Christian villages. This demographic tendency means that there are younger people around to participate in such organizations.[11]

Although the specific constellation of features in this case is unique, in my experience the individual elements of the story have been repeated in other Bulgarian environmental NGOs. I could cite similar examples of the various elements from other Bulgarian NGOs—of socialist-era sporting clubs registering as independent NGOs, perhaps with a different focus, since expeditions abroad are simply out of the question financially; of NGOs forming in order to avoid various aspects of the accounting structures of government agencies; of the importance of individuals; and so on. What is interesting in this case is, first, the number of these issues that are brought together in one community, and second, the fact that this is a small and relatively isolated community. This story of cavers and collective action thus captures key features of Bulgaria's NGO community and shows the pertinence of ethnography for evaluating ideas about nongovernmental organizations and civil society in the region. It also provides insight into the conditions under which such organizations develop.

Conclusion

The time: Late August 2000. The place: Rozhen, a high mountain meadow located on Bulgaria's one-time border with Ottoman Turkey.

For more than a century, this site has served as a meeting place for Rhodope residents, initially for families separated by the drawing of the international border after Bulgaria's independence. The border has long since been redrawn further south, but Rozhen still serves as a location for summer gatherings. This year, the periodic folk music festival has been adopted by the country's president and is advertised as "a fair for all Bulgarians." I arrive midmorning on the village bus with the singing ensemble from Zaburdo's cultural center. Leaving the bus, we hear the distinct sound of traditional Rhodope bagpipes coming from one of the stages. The crowd of people is impressive.

The girls, who number about a dozen, are scheduled to perform in the afternoon, so after locating the restaurant run by a village entrepreneur that will serve as our rendezvous point later, I go off to explore with my landlord's nine-year-old grandson. He has pleaded with his mother to let him accompany me, and the burden on me is slight because his grandfather is cooking at the restaurant and the boy spends part of the day with him. In addition to several stages for musical and dance performances, and a wooden tower topped with sheep bells that ring when the wind blows, the broad and freshly mowed hay meadow is covered with vendors selling food and various goods.

Our first stop is a tent decorated with a banner for the Bulgarian postal service, where we send my parents a postcard specially issued and stamped for the festival. My young companion then uses his pocket

*change to buy corn on the cob and an "air-conditioned" visor with
battery-operated fan. Much of what is for sale is typical fair fare—ice
cream, sunflower seeds, plastic toys—albeit with a Bulgarian twist.
There are rose oil products from the Valley of the Roses, halvah and
Turkish delight from the central Balkan town of Yablanitsa, and
bundles of dried medicinal plants such as Saint John's wort.
Restaurants serve grape brandy, Shopska salad (chopped tomatoes and
cucumbers topped with grated feta), and spit-roasted yearling lamb,
and vendors in one corner sell traditional handicrafts such as
embroidered slippers, textiles, and woodcarvings.*

*The booth that I am looking for is located next to that of the regional
historical museum. Two environmental NGOs—Green Balkans
and the Bulgarian Society for the Conservation of the Rhodope
Mountains—have convinced fair organizers to allocate them a booth
free of charge. It is the only one occupied by Bulgarian NGOs and
indeed the only apparent NGO presence. The booth is decorated with
photos, maps, and other information about a proposed protected area in
the Rhodope Mountains, and it is staffed by many enthusiastic young
people, including a Peace Corps volunteer assigned to Green Balkans.
They are collecting signatures from fairgoers on a petition in favor of
the nature park. An activist from one of the organizations later told me
that she was disappointed in their postcard and book sales, but she was
glad that Bulgaria's president had visited their booth and signed the
petition. Including those signatures from the fair, they have collected
more than 2,400 signatures in support of creating a park in the
Rhodope.*

For more than a decade, Bulgarian environmentalists both in and
out of the government have been discussing the potential designation of a large-scale protected area in the Rhodope Mountains.
Some discussions include the possibility of creating a park in
conjunction with Greece, in which case the park would be a
transboundary one, bridging what used to be called the "Iron Curtain." With the passage of a new law on protected areas in 1998,
and as Bulgaria's postsocialist restitutions of agricultural land and
forests draw to a close, thus clarifying issues of property ownership, conservation efforts seem to be accelerating. During 2001,
the United Nations Development Program (UNDP) approved

funding for conservation planning in the region, and Bulgaria's Ministry of Environment and Waters has authorized selected Bulgarian NGOs to draft proposals for parks in both the eastern and western Rhodope. The response to the petition drive described earlier indicates that popular support exists for this as well. Due to the mixed ownership status of the land and the fact that it is a relatively populated area, what is created will most likely be a "nature park" in Bulgarian terms, a "protected landscape" according to the internationally recognized protected area management categories of IUCN. Several established NGOs with histories of working in the region—Green Balkans, the Bulgarian Society for the Conservation of the Rhodope Mountains, the Wilderness Fund, and the Bulgarian Society for the Protection of Birds—are taking lead roles on this, although the prospect of substantial funding has also attracted the attention of other organizations. Some but not all of these groups have a history of working in the region, working on biodiversity conservation, or both. The question now seems to be "when" one or more new protected areas will be created in the Rhodope, rather than "if," and conservation advocates recognize that the populated nature of these mountains poses special challenges to these efforts.

In the Land of Orpheus has told the stories of two different sorts of communities interested in the same Rhodope Mountain landscape in an attempt to understand how the local realities of rural residents in the uncertain postsocialist context relate to global priorities concerning biodiversity conservation. The first is a geographically based village community located within these mountains, and the second is in essence an interest-based—and geographically dispersed—conservation community united in its concern about the region's landscape and biological diversity. This combination of detailed, "classic" ethnography of a rural mountain community and of conservation efforts and organizations has important lessons to offer. Previous chapters detailed the ways in which Rhodope residents are using natural resources, factors affecting the shape of this resource use, and the impact of the postsocialist context—national as well as local, political as well as economic—on Bulgarian environmental NGOs and their conservation efforts. This conclusion focuses on a few key questions regarding the intersection of rural communities, NGOs, and nature conservation that were raised in the introduction. First, what are the perspectives of the two communities regarding biodiversity, and can the NGOs and resource users find enough in common to work together? Second, what light does the material presented shed on assumptions about NGOs as linking agents in conservation efforts and as agents of civil society? And third, what do these conclusions mean for the future of the Rhodope?

Diverging or Converging Views of the Rhodope Landscape?

As Blaikie (1995) points out, different people may see and value the same landscape in different and socially constructed ways. Thus, one contribution that anthropologists can make to biodiversity conservation efforts is to identify the interests in and concerns about a given environment or landscape by the various stakeholders (Orlove and Brush 1996). The term "biodiversity" is not part of the everyday vocabulary of the typical Rhodope villager, but the local landscape—the pastures, agricultural fields, hay meadows, and forests, along with the plants and animals found there—is nevertheless critically important to them for the role that it plays in their livelihoods. In other words, the natural landscape is important to them largely as an economic resource. Earlier chapters described postsocialist economic conditions, in which many have lost their jobs and there has been a considerable decrease in the purchasing power of pensions and the remaining jobs, and detailed how people have responded by increasing their reliance on locally available natural resources. Residents of Zaburdo and other Rhodope villages have returned en masse to private agropastoral production on the restituted agricultural land in an effort to put food on the table and to earn some income or otherwise obtain products that they do not produce themselves. They grow potatoes, hay, and a few garden vegetables, and they raise cows, sheep, goats, and work animals. These activities provide subsistence goods, cash income, and a currency with which to engage in barter and other forms of informal exchange. Some people have also turned to or intensified their collection of wild mushrooms and herbs for commercial sale to gain cash income, and in a context of rising utility prices, locally available firewood is a critical resource for heating and cooking. Indeed, having potatoes in a storage shed, jars of home-canned fruit and vegetables in the pantry, and milk from a cow probably made life in Zaburdo less difficult and less uncertain during the worst days of the early 1997 economic crisis when stores closed and prices skyrocketed.

But when you talk to Zaburdo residents about the environment, about the place that they live, other things may emerge as well. Villagers mention the clean mountain air and the cold water that they drink out of mountain springs. This rural setting is described as quiet, calm, and crime-free when compared to Bulgaria's cities and the political disruption occurring there during my fieldwork. They are also aware of the region's tourism potential, and a couple of villagers indicated an interest in opening small family hotels or selling handicrafts to tourists. Beyond the limited hunting that occurs, some wildlife is considered problematic when it digs up agricultural crops or less frequently attacks sheep or chickens. But wild animals were never identified as a significant problem for agricultural activities according to my household survey data. Finally, it is useful to recall here that farming is not the first-choice occupation of all villagers, some of whom received specialized occupational training during the socialist era. Many would be happy to have a less physically demanding means of making a living should such an opportunity arise.

In contrast to the perspectives of rural residents, biodiversity and its protection are central to the vocabulary of Bulgarian conservation NGOs and government environmental officials. For them, a key reason for their interest in the Rhodope is its biodiversity and other natural values—the number and variety of species, the presence of rare and endemic species, the unique landforms, and so on. They also recognize, however, the populated character of the landscape, and their projects and programs often include components stemming from this. Sometimes this involves research on the relationships between rural communities and the environment, which is useful due to a scarcity of data on such matters. Other common efforts include the previously described public education campaigns and the promotion of economic development activities that are environmentally friendly, such as rural tourism and medicinal-plant cultivation. Conservation advocates also recognize that resource use by local people is not necessarily the only or even the greatest threat to the environment. In recent years, for example, Bulgarian NGOs have opposed on the basis of environmental concerns the development of new ski areas near the Rhodope towns of Smolyan and Velingrad, and they have also opposed hydroelectric power development on the Arda River. (Local residents may have opinions about these issues, particularly when they offer much-needed jobs, but the main force behind resource development is usually commercial interests or sometimes the government, not the rural communities.)

Comparing the perspectives of these two communities on biodiversity—or perhaps more to the point, on the Rhodope landscape—we can now consider whether NGOs and resource users have enough in common in terms of shared perspectives and concerns to work together. How might these perspectives affect relationships between rural residents and the government agencies and organizations promoting conservation? At first glance, these two stakeholder groups have different interests in the Rhodope landscape. Yet, the NGOs most centrally involved in Rhodope conservation efforts recognize that the landscape they wish to protect is a culturally modified one and that their efforts must take into account the economic futures of the rural communities located there. Meanwhile, some rural residents mention ecological concerns such as how herbs or mushrooms are harvested or the effect that logging might have on the water balance. The opinions of rural residents about creating a park will depend in large part on perceptions about what this means for their economic status and livelihoods. If they perceive it as a threat to their livelihoods, people may resist the idea of setting aside land as a park, but they may be supportive if they see the park as beneficial to their interests. While the potential for tension over issues like jobs is real, there is also some common ground regarding sustainable resource use that might provide the space for the NGOs to work together with local residents. This will not necessarily be easy. For example, in the context of the immediate economic needs facing rural villagers, it is possible that they might be put in a position to overharvest mushrooms (currently an open access resource), to overcut forests, and the like. But they should be able to understand each other enough to start a conversation.

NGOs, Biodiversity Conservation, and Civil Society

With the crumbling of communist-party rule in the late 1980s, conditions for an independent environmental sector in Bulgaria liberalized considerably, and the number of organizations has since grown significantly. These NGOs range from small, local-level groups to larger ones with national structures, and they work on a variety of issues including biodiversity conservation. The government's nature-protection activities are also receiving increased attention in the postsocialist period. These activities often are supported with financial and technical assistance from Western governments or international organizations. While the activities of independent organizations are not limited by the state in the same way that they were under socialism, modern legislation on these groups was only recently passed, and present-day Bulgarian NGOs operate under various constraints as well as conditions of uncertainty. Financial conditions in the country, lack of a tradition of philanthropy for such activities over the last 50 years, and questions about the legitimacy of this independent sector combine to make many organizations reliant on external donors for much of their financial support. Meanwhile, the ability of NGOs to influence public policymaking is constrained by their limited access to decision-making processes and the extent to which they are seen as legitimate players in environmental management and regulation. This is not to say that they are never viewed as having standing on environmental issues, but rather to point out that such access and standing are neither institutionalized nor guaranteed. Other factors affecting the activities of these organizations include the small sizes of the NGOs; urban or professional biases; the country's infrastructure (e.g., the banking system, telecommunications); and the preoccupation of many citizens with basic economic survival, which influences their level of participation in such groups.

One particularly relevant example of the way in which Bulgarian NGOs are affected by postsocialist conditions can be seen in their connections with government bodies. First, the involvement of government officials may lend credibility to the NGOs, given the lack of legitimacy and authority accorded to these independent organizations by some members of the Bulgarian public—or even a lack of familiarity with what such organizations are. Second, nongovernmental organizations may have easier access than some government agencies and institutions when it comes to obtaining money for conservation and sustainable development activities, due to the preference of some donors for NGOs. Consequently, some government employees in environmental fields (and other fields as well) have even created NGOs to get money for their projects. While such connections may call into question Western liberal assumptions about NGOs as part of civil society (as distinct from government), they may also be a fact of life for the Bulgarian environmental community in the postsocialist context.

Beyond this, what light does this study of the Bulgarian NGO community shed on debates about postsocialist "civil society"? I am less interested in holding Bulgarian environmental NGOs up to the light of western theory than I am in refining or critiquing that theory based on the Bulgarian reality. And the answer is basically that in several respects, the ideals of civil society do not fit very well Bulgarian environmental groups. First, it is clear that there was no immaculate conception of environmental NGOs in Bulgaria; in other words, these organizations did not just appear out of thin air with the crumbling of state socialism. Their history goes back to the beginnings of the modern Bulgarian nation-state, and more importantly, events and organizations in the socialist era set the stage for developments in the late 1980s and the 1990s. Second, NGO–government relationships in Bulgaria are messier than the ideals envisioned by advocates for civil society as a key to democracy. Due to the realities of funding in postsocialist Bulgaria, there is not infrequently an intertwining of government and nongovernmental sectors. While some might see this as problematic, such connections can also contribute to the legitimacy of these new institutions. In some cases, NGOs are little-known entities, or perhaps are viewed negatively because of their association with party politics around the time of the political changes. But when NGO officials are seen as being tied to a university, local government, or research institute at the Academy of Sciences, the positive reputation of these organizations may be transferred to the NGO as well. Third, it is important to recognize the impact of funding on the Bulgarian NGO community, and in particular the fact that funding concerns may influence the number of organizations created as well as the relations among them. One cannot simply count the number of NGOs as a measure of civil society. Funding can also impact the specific issues that the NGOs address, that is, they work on topics for which they can get funding. Finally, there is the question of what NGOs actually can achieve other than being a thorn in the side of the government. Their effectiveness in the policy arena depends on access to decision making and decision makers, and this access is only partly institutionalized in postsocialist Bulgaria.

Government conservation activities have also been affected by postsocialist conditions, and this, too, is important for understanding the context for NGO efforts. Environmental NGOs are not a replacement for sound environmental policies and enforceable laws, yet NGOs sometimes emerge in response to perceived shortcomings in government institutions and programs or a lack of government resources (Livernash 1992; Price 1994). Environmental issues played a role in and received attention around the beginning of the political changes in Bulgaria, and this carried over into the passage of an environmentally friendly constitution and a relatively strong Environmental Protection Act. More recently, however, such issues are receiving less attention from the National Assembly. It has been slow to update other environmental legislation and has approved weakening amendments to the environmental protection act. Beyond this, the institutional capacity of some government agencies for dealing with environmental issues continues to be weak. This capacity extends to the reliance on external donors by the cash-strapped government for much of its funding for environmental projects, through such entities as the GEF Project and the Bulgarian-Swiss Biodiversity Conservation Program. Sometimes there is a lack of resources for the enforcement of environmental regulations, and even limited procedures for financial administration of the funds received. Indeed, given the current fiscal situation for Bulgarian government bodies, working through an NGO is sometimes an efficient means for achieving environmental objectives or conducting ecological research (see Wells and Williams 1998). In some cases, the NGOs may be seen to be more effective at putting such funds to environmental purposes than government institutions with their weak financial management structures. The latter prompted both the Bulgarian-Swiss program and Trigrad's caving club to register as nongovernmental organizations in an effort to direct funds more efficiently to environmental projects.

The issues to consider at this juncture, given these constraints, are first, the role of NGOs generally as an institutional mechanism that mediates between the local, national, and international levels in conservation efforts. and second, the more specific influence of environmental organizations on natural resource use and management in rural areas such as the Rhodope. In an article on Bulgarian politics in the postsocialist period, Daskalov writes that NGOs generally have been of "very little significance. Their activities are limited at best; at worst they are entirely geared towards receiving aid from outside agencies and foundations and do little except prepare grant applications and reports" (1998:18–19). Jancar-Webster (1998), meanwhile, talks about the transformation of the environmental movement in Central and Eastern Europe from a mobilizing movement for popular protest to goal-oriented professional organizations. While acknowledging the good work of such professionalized organizations, such as their participation in environmental decision making, she also raises concerns about the implications of this transformation for environmental movements in the region. These concerns include the potential for a gap to develop between grassroots organizations and the well-funded and connected groups and the lack of incentives for ordinary citizens to participate in environmental movements as they struggle with the tasks of daily living. The material presented here concerning the conservation activities of Bulgarian NGOs shows, however, a more complex picture when one examines the flows of information, ideas, people, and money, and the impact of these flows in particular places and times.

Even with these constraints and the movement's limited size, the activities of environmental NGOs in Bulgaria merit more attention than the observations of the scholars cited above might suggest, in part due to the situation with the country's government environmental efforts (e.g., limited enforcement capacity, reliance on foreign funding). Besides funds for conservation activities, international sources have provided Bulgarian groups with such resources as information, training, and the opportunity to see how NGOs operate in other countries. Given the frequent changes in government, these ties between NGOs and their outside donors may be longer lasting than those between the government and its outside donors (see also Wedel 1998a:118). The potential for reliance on foreign donors to have an undue influence on the activities of Bulgarian NGOs and the Bulgarian government certainly warrants attention as the situation unfolds. Examples of this influence include the University Rescue Squad's focus on homeless children and orphanages instead of on activities that use the technical skills of its members, and the Foundation for Environmental Education and Training's seminar on ecotourism when it had no previous experience with the issue. Yet, positive things can be seen as resulting from these contacts and resources. While anthropologists often focus their attention on how individuals and households are surviving the postsocialist period, the organizations, too, must survive postsocialism if they are to have an impact on Bulgaria's environmental future (Bridger 1998; Pine and Bridger 1998:9), and foreign funding has contributed to this. One factor clearly influencing the degree to which the NGOs were affected by the economic and political chaos of early 1997, for example, was the extent of their international connections and funding.

The financial and technical support gained through such international contacts have helped Bulgarian organizations, as well as the Bulgarian government, to collect environmental data and implement conservation projects on the ground. Some NGOs have also been active in the policy arena, commenting on legislation and advocating for the creation of protected areas. The Bulgarian Society for the Protection of Birds proposed the creation of the Poda Lagoon nature reserve, for example, and is now involved in its management. The habitat maintenance and feeding activities of BSPB as well as Green Balkans in places that include the eastern Rhodope have contributed to an increase in the nesting populations of some bird species. Contacts with outside organizations have also given the Bulgarian NGOs access to information about what is happening in environmental circles in other parts of the world. This is seen, for example, in the Wilderness Fund's consultation with IUCN protected-area experts in developing its comments on Bulgaria's new protected-areas law. The NGOs and bilateral conservation projects have also employed scientific professionals in their areas of expertise during this time of economic crisis in Bulgaria, and in so doing provided them an alternative to migration to the West or taking jobs outside their fields.

While Jancar-Webster (1998:56) raises concerns about the professionalization of environmental NGOs in Central and Eastern Europe and the potential to lose touch with the grassroots as a result, evidence in the Bulgarian case suggests a more varied community and a more complex set of relations between groups at different levels and between the NGOs and rural communities. The mass mobilization of the population around environmental issues in 1989–1990 has declined, and most organizations are small, indeed perhaps smaller in many cases than the appellation "NGO" might suggest. Yet, appreciable variety still exists in the Bulgarian NGO community in terms of such things as location, structure, size, and the constitution of their memberships. A few well-funded organizations have regular professional staffs, some hire people only on a project basis, and others operate almost entirely on a volunteer basis, with most funds going to nonpersonnel project expenses. Several groups have struggled on shoestring budgets for nearly a decade. Clearly, some professionally oriented, urban-based groups are more biologically focused and promote their expert status. But contrast this with the Bulgarian Society for the Conservation of the Rhodope Mountains, a regionally oriented group that is concerned about natural and *cultural* heritage of the mountains, arguing that you cannot have one without the other. While it is more common for Bulgarian environmental NGOs to be located in urban areas, there are also groups in more rural settings concerned about environmental and resource-use issues. Of course, these rural organizations may address these issues from the perspective of sustainable communities or economic development rather than focusing on technical, biodiversity-focused concerns.

Beyond this, many of the conservation NGOs I contacted have local connections of some kind, try to incorporate community concerns into their conservation activities, or both. The Wilderness Fund might be characterized as a more professional group given its small size and the scientific orientation of many members. Yet, some of its projects involve or at least communicate with and address the problems of rural communities, and its connections to these communities have employed ties through former students of members working in communities, ties due to employment-related trips to research sites, and the family connections of NGO members to rural communities. Other organizations provided similar examples of ways in which such informal ties were employed in their activities. The Bulgarian Society for the Conservation of the Rhodope Mountains and the Bulgarian Society for the Protection of Birds are somewhat similar in the professional constitution of their national leaderships, but they also have some local structures—including in the Rhodope—and ordinary people as members. (As I write this, the current president of BSCRM is a retired schoolteacher living in and from a small Rhodope town; in the 1980s she was active in the local caving club.) This structure can be useful in the linking role of NGOs. An official of the Plovdiv branch of BSCRM described a role for that group in mediating between local and national levels, for example, in saying that that people from the rural branches can call someone in that city for assistance in contacting the correct authorities when a problem arises. Thus, some degree of professionalization is not necessarily bad. As Guha (1997:26–27) points out in discussing environmental activism in India, one advantage of having key leadership roles in an organization filled by activists who are from the region but not directly involved in agricultural or pastoral production (e.g., engineer, social worker, labor organizer) is being able to draw upon people with the "experience and education to negotiate the politics of protest" (Guha 1997:27). Such is the membership composition of the Bulgarian Society for the Conservation of the Rhodope. Finally, the national-level groups can also, in principle, play an intermediary role in mediating between the global and the local through such things as communicating problems originating at the local level to regional or national entities (facilitated by the connections just described) and also sharing their experience with local groups, including on matters of fundraising. As well, perhaps they have a better understanding of local conditions—for living as well as working—than an international donor might have. Since most international conservation organizations working in Bulgaria typically operate through national or regional NGOs, such a role is possible.

Finally, what can be said about the impact of NGOs on rural communities and on local-level resource use and management in the Rhodope? In the last decade, a few environmental organizations have been created in the Rhodope, and some other NGOs have engaged in conservation projects there in recognition of the region's biological diversity. Often these projects have focused on the collection of environmental data, environmental education, and the promotion of ecological tourism. Projects that directly address the kind of resource uses that were described for Zaburdo were comparatively less frequent, but they do exist, nonetheless. A project to promote the cultivation of medicinal plants was not successful, although factors beyond the local community led in large part to this failure. In other cases, informational brochures or posters have been prepared about medicinal-plant or mushroom collection, at least in part to promote more sustainable resource use. Green Balkans is trying to reach out to rural communities through its tourism and sustainable development project in the southern Rhodope, although in this case the model being used is to assist the communities in creating their own local organizations. This is similar to the linking role that Carroll (1992) describes for what he calls intermediary organizations. Such a role has seen relatively limited application to date in Bulgaria, although factors potentially contributing to this are the relative newness of the national-level NGOs, on the one hand, and limited number of rural organizations, on the other. One problem in some rural areas is the absence of a local group with which to work, which makes such an intermediary role difficult. In other cases, however, such as Trigrad, there are local groups.

Can NGOs play the promised role as linking agents in conservation activities? Is what they are doing affecting local resource users and resource use? The bottom line is that NGOs do play a role in linking global biodiversity concerns with local realities. Along with policy efforts aimed at creating some sort of protected area in the Rhodope and the collection of ecological data, NGOs are doing several things intended to help rural populations use resources in a more sustainable manner. Beyond ecotourism, these activities include promoting more sustainable techniques for harvesting mushrooms, the raising of local varieties of domesticated animals, and the cultivation of medicinal plants. That said, the NGOs are not perfect. As Vivian (1994) writes, they are no "magic bullets." Their projects are not always successful, in some cases for reasons that are not within their control. These organizations do not have a lot of experience as organizations—a bit over a decade at most—and they are working in a changeable situation with limited financial resources and sometimes questions about their legitimacy as stakeholders or their effectiveness as political actors. Yet the projects can be useful, and even the failures can be learning experiences. In the Bulgarian context, NGOs can be more flexible, stable, and issue-focused, particularly when compared to a poorly funded government distracted by the latest political crisis.

NGOs, Rural Residents, and the Future of the Land of Orpheus

It is now time to draw together the strands of this study and to consider their implications for the future of the Rhodope with regard to conservation and sustainable development. (In so doing, I take the presence of rural communities as a given and focus on the role of NGOs.) Foretelling the future is difficult, of course, due to the number and importance of the "ifs" involved and the difficulties of predicting what will happen in Bulgaria. If NGOs had an easier time with fundraising and were not so dependent on international grants, they might spend less time searching for funding and more time on the ground in the Rhodope landscape. This could provide an opportunity for more effective work with rural communities or with a larger number of communities. If Bulgarian citizens were better off financially and thus had more time, money, or both to share with NGOs, the groups might have broader membership bases. If the village economic situation was better, there could be less pressure on nontimber forest products. Or, if the village school closes, families with children are likely to move to town leading to further depopulation of the landscape. And what of the effect of joining the European Union on agricultural or environmental policy?

As we have seen, NGOs currently have a somewhat limited direct impact on resource use and management in rural Rhodope communities, in part due to the political and economic contexts in which they operate. Their numbers and financing are limited, and the Rhodope Mountains cover a large territory. Yet, there are environmental organizations working in rural communities, and people from the Rhodope are members of some regional and national organizations. These organizations are involved in such things as the management of protected areas and concrete efforts to support particular species. In part, these roles are made more important by the weakness of government infrastructure for environmental management.

In the short term, NGOs will likely play a role in protected-area designation for the Rhodope, and to the extent that the organizations involved are also groups with a Rhodope presence, they may influence the way in which subsequent management regulations address issues concerning rural communities. The expertise of NGOs and their commitment to environmental issues is recognized, for example, in the desire to involve them in the UNDP Rhodope project. Conversations with activists of the Green Balkans, who have been visiting rural municipalities as part of park planning efforts, also indicate that they are now often recognized in rural communities as legitimate actors. In addition, their work to promote rural tourism as an activity with a potential economic benefit to rural residents may raise awareness among rural residents of the organizations more generally (cf. Close 1998:58). This is in some ways related to the lack of familiarity with such independent groups. Some NGOs have indicated an interest in helping rural communities diversify their agricultural production beyond potatoes and tobacco (e.g., berries) and also in marketing that production as having come from an ecologically clean area. As I left the field, a couple of NGOs were considering the ways in which they might contribute to forest management practices as the forests are reprivatized, and this is another area in which there is a potential for NGOs to affect local-level resource management. Thus, although the community-level impacts of NGOs were somewhat limited in the 1990s, they are already connected in personnel as well as through interest in the landscape, and there are grounds to believe that they will have a role in future resource use in the land of Orpheus.

Notes

Introduction

1. One still sees references to Orpheus in the region, however. A children's music festival in Smolyan, a community's *chitalishte* (cultural center), and a village-based NGO are named after him; the cover of a recent tourist brochure identifies the Rhodope as "the mountains of Orpheus"; and there is a statue of him near Hotel Orfei at the Pamporovo resort.

2. "Socialist" and "communist" are used interchangeably here to refer to the period in Bulgaria between 1944 and 1989. The term "post-socialist" is used for the post-1989 period, although it must be recognized that during some parts of the last decade the Bulgarian Socialist Party—a reformed successor to the communist party—has controlled the national government.

3. Before embarking on the journey described here, my knowledge of Eastern Europe came primarily from my high-school German teacher, who insisted that we learn something about the region. As I considered where to do my study of local resource use, environmental NGOs, and biodiversity conservation, I was drawn to the region in part due to these high-school experiences and a vague hope that my rusty German-language skills might somehow come in handy. I was also intrigued by the question of what changing property relations and a relaxation of controls on independent environmental activity meant for conservation and resource use and also by the fact that the region—and particularly Bulgaria—had been little studied by anthropologists, due to socialist-era restrictions on researchers, particularly those wanting to live in rural areas.

4. Northern typically refers to the more "developed" countries (and the organizations therein) of the U.S., Canada and Western Europe,

279

while southern refers to countries (and the organization therein) located to the south that tend as a gross generalization to be the recipients of aid rather than donors.

5. This section focuses on literature in English by Western-trained social scientists. Many of these countries have long-standing traditions of folklore, ethnography, and sociology, and their practitioners have produced studies of relevance to anthropologists—provided that they were able to do so within the ideological constraints of the political regime in which they were working in the past and the economic constraints of the present. Many of these studies are published in local languages, and they vary in the attention paid to issues of concern to anthropologists and the use of ethnographic research methods.

6. I am referring to scholars such as folklorist Klaus Roth (1990), folklorist and anthropologist Carol Silverman (1983, 1984, 1986), anthropologist Eleanor Smollett (1980, 1989), and sociologist Roger Whitaker (1979).

7. In writing this, I keep thinking that I must be missing something, but a scan of a database of U.S. dissertations—where one might expect to see the first results of such long-term projects—turned up only my 1999 dissertation as being obviously based on long-term rural fieldwork, the only such work since the studies of Gerald Creed (e.g., 1998) and ethnomusicologist Donna Buchanan (1991). While representing scholarship in only one country, it is indicative of the general trend.

8. Examples include Bridger and Pine 1998; Creed 1991, 1993, 1998; Hann 1980, 1993a, 1993c; Kideckel 1993b, 1995a, 1995b; Lampland 1995; Nagengast 1991; Roth 1990; Sampson 1993, 1995; Smollett 1980, 1989; Verdery 1993, 1996; also, for Russia, see Grant 1995; Humphrey 1998.

9. Book-length studies focused predominantly on the postsocialist period are relatively less common; however, a number of useful edited volumes have been published. They include Berdahl et al. (2000); Burawoy and Verdery (1999); De Soto and Dudwick (2000); Hann (2002); Kideckel (1995b); and Leonard and Kaneff (2002).

10. The exchange rate was initially set at 1000 leva = 1 German mark; following redenomination in July 1999, it became 1 lev = 1 mark; and with the introduction of the Euro, 1 lev = 0.51 Euro.

11. Since I originally wrote these words, some of my Bulgarian friends have gone to the United States with so-called green cards and another is working in Greece. Many other Bulgarians I know have applied for green cards in the hope of finding a way out.

12. His name was not on the list of candidates, however.

13. Initially elected on a UDF ticket in 1996, he ran for reelection in 2001 as an independent candidate, albeit with UDF endorsement as well as that of the king.

14. In Sofia, I rented a room in the apartment of a city resident, rode public transport, shopped in neighborhood markets, and socialized primarily with Bulgarians, thereby gaining a window into urban life more generally.

15. Most of my time was spent with women or in mixed groups; however, some of the activities listed—for example, livestock herding and hunting—are predominately male ones. As one might expect, the people I accompanied were protective of me on these occasions, but I did not encounter resistance to my tagging along, particularly when it became clear that I could keep up with the group, did not complain about the weather or distance, and so on.

1. Bulgarian Environmental NGOs and Nature Conservation

1. The founding groups were the Bulgarian Botanical Society (established in 1923), the Bulgarian Nature Research Society, the Society of Forestry Academicians (est. 1928–1930), the Society of Bulgarian Foresters (est. 1909), the Bulgarian Hunting Organization, the Bulgarian Fishing Union, the Bulgarian Caving Society (est. 1928), the Bulgarian Geological Society (est. 1925), the Bulgarian Tourist Society (est. 1895), and the Junior Tourist Union (est. 1913) (Suiuz za zashtita na prirodata 1995).

2. A 1993 catalogue of Bulgarian environmental NGOs confirms that the Student's Environmental Protection Club at the Forestry Institute was created in 1977 (Penchovska et al. 1993:49).

3. These figures predate the reclassifications and border adjustments necessitated by the 1998 protected-areas law, from which the dust has yet to settle. The most important change is that only three parks remain classified as national parks, with most if not all of the remainder being reclassified as nature parks. In a related development, 23,000 ha have been removed from Rila National Park in order to restore this territory to the Rila Monastery under Bulgaria's forest restitution legislation. As part of this deal, the removed territory was designated a new nature park.

4. While the 1991 constitution declares parks and reserves of national significance to be state property, it also says that when property is expropriated to meet state needs, suitable compensation must be paid in advance (translated in Flanz 1992:89, 90).

5. PHARE is a large aid program established by the European Union. The initials stand for Poland-Hungary Aid for Restructuring the Economy, reflecting the program's initial focus on those countries.

6. This project was part of the U.S. government's contribution to the Global Environmental Facility (GEF). The GEF was set up as a pilot project in 1991 and restructured in 1994. Managed through the United

Nations Development Program, the United Nations Environmental Program, and the World Bank, it provides grants and concessional funding to countries for projects and programs that protect the global environment and promote sustainable economic growth. The GEF is the funding mechanism for the Convention on Biological Diversity and the Framework Convention on Climate Change (GEF Secretariat 1999).

7. Smaller conservation and sustainable development projects in rural Bulgaria with Swiss sponsorship include a sustainable forestry program, along with ecological agriculture and rural tourism activities in the Balkan Mountains.

8. Members of the Bulgarian Society for the Protection of Birds and Green Balkans have continued these feeding efforts. This activity has contributed to the comeback of the globally threatened black vulture in Bulgaria and an increase in the number of nesting pairs of griffon vultures (Peev et al. 1995:53).

9. Because being designated as such involves more paperwork, some environmental NGOs that clearly benefit more than just their members are registering as organizations with private purposes.

10. In focusing on nature conservation and nature conservation NGOs, I do not intend to downplay the importance or seriousness of other environmental problems in Bulgaria. As mentioned earlier, some of the earliest environmental protests concerned air pollution from a plant in Romania, and Bulgaria has its own share of pollution problems from mining, chemical, and petroleum-processing facilities. Nuclear power is another contentious issue, in the aftermath of the Chernobyl disaster, with the Russian-designed reactors at Kozlodui providing about 40 percent of the nation's electricity. Here, too, there are examples of government officials downplaying risks while NGOs and some local residents raise concerns about plant safety and waste disposal (e.g., Konstantinov 1995; Nikolova 1998).

11. Koulov (1998:210) observes that as conditions worsened, public interest in environmental affairs declined, and active participants in Burgas-based NGOs were reduced to people professionally employed in environmental fields (see also Nikolov 1992). Some people may join environmental organizations or continue their earlier moral support even if they are unable to be active. Such is the case with one of my young Bulgarian friends, who frequently discusses her environmental interests but is too busy earning a living to participate significantly. Pickles et al. suggest that these economic conditions, in which people struggle to make ends meet, "have dislodged environmental issues— although . . . not environmental *concerns*—from the public agenda" (1998:245, emphasis in the original).

12. One potential Bulgarian funding source for biodiversity projects

is the National Trust EcoFund that was created as the result of a debt-for-the-environment swap between the Bulgarian and Swiss governments in 1995. (This was the second such agreement in Eastern Europe, after one made by Poland.) Although biodiversity conservation is one of the fund's four priority areas and NGOs are among the actors eligible to apply for project funds, very little money had gone to this purpose when I visited the fund's offices in 1998, and none had gone to NGOs. The NGO officials I spoke with did not consider the EcoFund a particularly important funding source (National Trust EcoFund 1998).

13. This line of inquiry was suggested by Gerald Creed during a conversation in the midst of the crisis.

14. A Bulgarian friend employed in the tourism industry related that at least one Western European country issued a travel advisory for Bulgaria following the 1997 crisis, and she consequently lost some tour business.

2. Landscape, Community, and Economic History in the Central Rhodope

1. About 55 percent of the residents of the Smolyan Region (*okrug*) and 19 percent of the Chepelare Municipality (*obshtina*) identified themselves as Muslims on Bulgaria's 1992 census, compared to 13 percent for the country as a whole. Similarly, 7 percent of the residents of the Smolyan Region and less than 1 percent of the Chepelare Municipality identified their ethnicity as Turkish, compared to about 9 percent for all Bulgaria. Many non-Turkish Muslims in the region—the difference between these two sets of figures—constitute a group often identified as Pomaks or Bulgarian (-speaking) Muslims, although the census included no designation for them. Nearly all of the Muslims in the Chepelare Municipality, in which the study village is located, fall into this category. Roma or gypsies—who may be Christian or Muslim—accounted for less than 0.5 percent of the regional population, while nationwide they made up about 4 percent of the total according to official statistics (National Statistical Institute 1996b:133–139; Zhelyazkova 1998:172–173).

2. Disguising the village name seems pointless, since I was officially registered there as a resident and it is well known among officials and others in the municipal center that it is "my" village. In addition, the village's identity is obvious from its characteristics in terms of size, location, and religious background. When discussing individuals, however, I do use pseudonyms.

3. In so doing, the research upon which this book is based falls within the analytical tradition of political ecology, which emphasizes the importance of considering local social practices and environmental

conditions, as well as the larger political economy in which they are embedded (Blaikie and Brookfield 1987; Cole and Wolf 1974; Greenberg and Park 1994; Peet and Watts 1996; Wolf 1972; cf. Colson and Kottak 1996 and Pelto and DeWalt 1985, who also raise concern about the larger context). I take political ecology as a general framework, suggesting attention to both the local situation and the larger context without giving priority to a particular explanatory factor (cf. Vayda and Walters 1999).

4. Given their location straddling the so-called Iron Curtain, another reason for this is that economic activities and to some extent even human presence were closely controlled in border regions during the socialist era.

5. Bulgaria's system of territorial administration has alternated between larger and smaller units since independence in 1878. Since no single system for translating the terms exists, the usage here is explained. During my 1996–1998 research, there were nine *oblasti* (translated here as districts), which were subdivided into about 255 *obshtini* (municipalities). The *oblasti* were established in 1987 to replace about 28 *okruzi* (sometimes referred to as *regioni* and translated here as regions). Within each *obshtina*, there are *kmetstva* or mayoral administrations (National Statistical Institute 1996a:334; Wyzan 1997). Zaburdo is in the Chepelare Municipality, the Smolyan Region, and the Plovdiv District. The system has since changed again, returning regions to greater significance.

6. More recently (2001), there is renewed interest among NGOs in Bulgaria and Greece in creating a transboundary park that encompasses the Rhodope in both countries.

7. Before collectivization, residents made extensive use of outbuildings located near their fields.

8. With few births in the last decade, there have not been enough children to make up classes in the village school in some years, and its director is concerned that it will soon be closed. This may prompt more young families to move away in search of educational opportunities for their children and would also eliminate several nonagricultural jobs.

9. Residents sometimes complain, however, about the lack of a pharmacy and a store selling commercial agricultural inputs such as fertilizer and concentrated livestock feed.

10. A partial exception is a few retirees who may go by their Muslim names informally while having Bulgarian names on paper. It is also worth noting that the rebuilding is necessary because religious facilities often fell into disrepair with socialist-era restrictions on religious practice.

11. Bulgaria's 1991 Constitution guarantees religious freedom, but identifies Eastern Orthodoxy as the "traditional religion" of the Bulgarian people. According to the 1992 census, 85 percent of the population is

Eastern Orthodox, other Christian affiliations make up about 1 percent, and 13 percent identified themselves as Muslim (Zhelyazkova 1998:173).

12. Subsequent name-changing campaigns appear to have encountered less resistance.

13. In Zaburdo, for example, a common practice was to give daughters one plot, referred to as *miraz*, as their land inheritance at or around the time of marriage. The remaining land was divided among the sons or other male heirs after the death of a landowner.

14. Some livestock owners subsequently took their flocks north to lowland areas around Plovdiv, but the scale of the transhumance was smaller and this did not address the loss of markets.

15. This occurred during a period in Bulgaria's history when the government was controlled by a peasant party, the Bulgarian Agricultural National Union (see Bell 1977). While land ownership remained in private hands, an economic development strategy based on co-ops was encouraged in an effort to counteract the fragmented nature of property holdings and improve conditions in the rural areas from which the party drew its support. This built upon earlier trends toward the creation of cooperative institutions, including a 1907 law on cooperative associations (Crampton 1987, 1997; Lampe 1986).

16. Potato late blight led to the Irish potato famine 150 years ago and has been experiencing a worldwide resurgence since the early 1980s (Fry and Goodwin 1997).

17. In some years, brigades of young people from outside the community provided additional labor, thus helping ease the labor shortage. They were required to do labor service during school vacations.

18. Bulgarians typically measure land area in decares. A decare is 1,000 square meters, which equals one-tenth of a hectare and about four-tenths of an acre. The predominance of smallholdings is reflected in another measure sometimes encountered: One *ar* (pl., *ara*) equals one-tenth of a decare or 100 square meters.

19. That such high percentages of these products came from a relatively limited area of land is in some sense evidence of the inefficiencies of the collective system. So too is the practice of feeding bread to livestock.

20. For brief periods in 1992 and 1996–1997, the facility was again used as a sewing workshop under outside private management but with a much smaller workforce. When I arrived in December 1996, winter jackets were being produced and later athletic suits. After a month full of rumors of impending closure, the facility shut its doors once again. In the end some workers received a jacket instead of their final monthly wages—which would have only been worth $5 to $10 as a result of the hyperinflation at the time.

3. Postsocialist Strategies of Mountain Agriculture

1. Chepelare has the closest weather station to Zaburdo for which continuous, long-term data are available. This town is located about 15 kilometers southeast of the village and about 100 meters lower in elevation. The Institute for Meteorology and Hydrology of the Bulgarian Academy of Sciences provided me with these data free of charge for my scholarly purposes; however, within a given month, 10 percent of the data points were scrambled to prevent their commercial use.

2. Beyond periodic droughts, domestic water-supply problems are exacerbated by problems with aging infrastructure, lack of water meters, and use of drinking water for irrigating garden plots.

3. In talking about interwar Dragalevtsi, for example, Sanders writes: "At first I could scarcely believe my eyes when I was shown how the peasants distinguished their fields. Taking some landmark such as a tree, a bend in the road, or a stream, the peasants marked off their strips by small piles of stones. Fences and hedgerows were impracticable because the strips were too small to justify the expense of enclosure" (1949:56).

4. Land inheritance in Bulgaria is typically partible, with holdings being divided among the children of a landowner, and both men and women may inherit. Although some authors (Lampe 1986) suggest that this is a straightforward equal division, the picture painted by Sanders (1949:55) is more complex. He reports that sons in Dragalevtsi in principle received twice as much land as daughters, although daughters sometimes forfeited their claims and favorite sons occasionally received larger shares. He also notes that the division of patrimony was the cause for the most serious conflicts in the village and that such conflicts periodically resulted in lawsuits.

5. Private agricultural cooperatives have been formed in other Bulgarian communities. According to Meurs and Spreeuw (1997:282), formally registered agricultural co-ops had been established in 67 of 99 villages included in their 1994 nationwide survey, the majority of them in grain-producing regions. Some Zaburdo residents express an interest in cooperative production, and the possibility of creating a co-op was occasionally raised at village assemblies. Others oppose cooperatives, however, and these discussions never got very far.

6. Such patterns have been described elsewhere. Netting (1981:11, 18) observes that barns and small huts near fields distant from the Swiss mountain village of Törbel reduce the time and labor needed to transport people, hay, and manure. Stevens (1993:168) similarly describes a practice in highland Nepal whereby hay is stored in outbuildings near the fields and animals are moved to it. He also notes that this contributes to manuring of the fields.

7. *Banitsa* is a ubiquitous Bulgarian pastry typically prepared using thin strudel-type pastry leaves and filled with cheese, or with rice and eggs, or with spinach, leeks, pumpkin, walnuts, or even potatoes.

8. The same can be said for domestic fruit production. Nearly half the households surveyed own fruit trees such as apples (32 percent), pears (16 percent), or cherries (15 percent), or they produce fruits like strawberries (17 percent).

9. I observed no widespread harvesting of hay from communal pastures, although during the 2000 drought forestry employees returned to the village each evening with their van filled with grass that they had cut in the forest. Their sun-parched hay meadows promised low yields, but their animals still had to eat.

10. In response to specific questioning, more than three-quarters of surveyed households reported that animals such as wild boar and roe deer caused problems for their crop production by digging up fields or eating produce, and 70 percent of them took precautions such as fences and scarecrows. Yet, animal damage was never identified as the most significant when they were questioned more generally about problems with household agriculture.

11. *Blizki*—short for *blizki hora*, literally, "close people"—play key roles in social and economic life in the village and beyond. An ego-centered group, this category of people can include relatives by blood or marriage as well as neighbors, schoolmates, coworkers, or fellow members of a political party. When asked with whom they exchanged agricultural labor, villagers often responded "our *blizki*." *Blizki* can also be of assistance beyond cooperation in agriculture. In Bulgaria today, as well as in the past, *vruzki* or personal connections, often mediated through networks of *blizki*, can be or are at least are perceived to be the key to such things as obtaining scarce goods, getting telephone service installed in a timely matter, gaining access to a good doctor, or finding a reliable mechanic who will not overcharge or otherwise cheat you.

12. Indeed, day wages are often quoted as being with or without meals, reflecting meal preparation costs.

13. I have followed Bulgarian convention with regard to the spelling of family names. Names of men typically end with a consonant, female names end with the letter "a," and plural names (e.g. a multi-person family) end with the letter "i."

14. A similar pattern of private ownership of intensively exploited fields and meadows and communal management of more extensively used pastures is described by Netting (1976) for a village in the Swiss Alps.

15. The importance of concentrated forage is also seen in the efforts of the owners of a small cooperative dairy in another Rhodope village. With

financial and technical assistance from a Swiss foundation, a small group of people started this dairy. They subsequently approached the foundation about establishing an agricultural co-op in order to provide local milk producers with reasonably priced forage. As a larger purchaser, the co-op would get better fodder prices than individuals and then pass cost savings onto farmers. In this way, the dairy hoped that milk production would be cost-effective for villagers and that they would continue to supply the dairy. Otherwise it was feared that farmers would get rid of their cows, and the dairy would have no milk to process.

16. A few people truck their calves to a neighboring village to go out with a paid herder in the summer.

17. In one neighboring village, a young man collects sheep from village households in the morning and returns them in the evening for a monthly fee. Under another arrangement seen in some Rhodope locations, a paid herder takes animals out daily from a herding facility located high in the mountains where the animals spend the entire summer, as was done during private times. The owners take turns showing up a few times a season to milk the sheep and then process the resulting milk into cheese and butter. The amount of milk an owner gets is based on a measurement of the milk produced by his or her sheep at the beginning of the season.

18. With Bulgaria's recent forest restitution some of these forest lands have been returned to private ownership, and these products are legally the property of the landowner. It is too early to tell whether they will try to impose access regulations and enforce their property rights (see Cellarius 2002).

19. Perhaps these are the Bulgarian equivalent of the candy sales and recycling drives I grew up with in U.S. public schools.

20. Mushrooms are perhaps even more weather-dependent than agriculture. There were almost no mushrooms in 2000 due to heat and drought, while 2001 produced a bumper crop, and it was not unusual for collecting groups to come home with mushrooms worth 70 or 80 leva (after the currency revaluation of July 1999; U.S.$32–38). Coming as it did after a year with no mushrooms and a poor potato crop, several people remarked that if it hadn't been for the mushrooms, and specifically mushroom sales, many village households would have been hungry that summer.

21. Nearly two-thirds of mushroom sellers interviewed sometimes also eat wild-collected mushrooms. Those who do not often cite concerns about poisonous mushrooms, and their concerns are likely reinforced by periodic media reports about poisoning incidents.

22. Mushroom prices vary over the season, starting high, falling as supply increases, and rising at the end as amounts dwindle. At the time

of these 1997 interviews, first-quality king bolete sold for 4,000 leva or about $2.22 a kilogram; second quality sold for 1,000 leva or about $0.56 per kilogram; and third quality sold for 500 leva or $0.28 per kilogram. Chanterelles sold for 1,000 leva or $0.56 per kilogram.

23. Following Bulgaria's forest restitution in 2000–2001, many forests around the village have returned to private ownership. It is too early to see the results of the privatization, however, and thus I do not address it in this book.

4. Making Ends Meet

1. Unemployment is difficult to quantify in this context. Lack of nonagricultural job opportunities is clearly a problem in rural communities, particularly in the winter when extensive agricultural activities are not possible. For example, data from the December 1992 census indicate that 40 percent of Zaburdo's working-age population was out of work, compared to 19 percent for the Smolyan Region (National Statistical Institute 1996b:238, 247). Yet, most villagers are busy with household agropastoral activities or mushroom collection during the summer, and they produce a considerable share of their annual food supply plus some cash through these activities.

2. Once the dust settles from the restitution of most forests in the village territory to their presocialist owners, tree planting may resume (reforestation is required by Bulgaria's law on forests). The number of jobs may, however, be lower when market principles rather than socialist employment policies govern hiring levels.

3. Especially before the establishment of the currency board and fixing of Bulgaria's currency to the German mark in 1997, hard currency, such as the U.S. dollar or German mark, was important for its ability to hold value relative to the sometimes rapidly deflating Bulgarian lev.

4. Another reason for this may be that there were not enough children their son's age to make a class in the village school—a result of declining birth rates—and with his mom working in town, he can attend school there.

5. Humphrey (1998:chap. 9) discusses the importance of pensions for postsocialist economic survival in rural Russia. In the case she describes, pensions have particular significance because they are more often paid in cash, which is valued for its instant convertibility, while workers may be paid in less-convertible goods.

6. An example of small-scale exchange was described to me by a friend from another Rhodope village. Residents of a village up the road (elevation-wise) from his village would come to his community to trade their potatoes for items that they could not produce, such as beans or

walnuts. He explained that in this fashion it was possible to obtain goods that were not produced in a particular village and also were not necessarily available through the formal state distribution system. Creed (personal communication, 1998) similarly reports that in the foothill village in northwestern Bulgaria where he worked in the late 1980s, some trade of produce from private production took place, albeit more commonly among a village's residents, rather than between residents of different villages.

7. My conversation with the mayor cited at the beginning of this section took place in late 1996 before the economic collapse. On another occasion, a man explained that during his tenure as mayor of another village in the early 1990s, he had arranged, through some connections, for the exchange of potatoes from that village for high-quality sugar from Yugoslavia, an exchange with which his constituents were said to have been very happy. Such exchanges also took place with the fall 1997 potato crop, which is to say after the implementation of the currency board in Bulgaria and substantial stabilization of the economy.

8. Another consideration may be whether a given product is available for purchase with cash or is only available by bartering for it with potatoes. If one can only obtain a particular, desired item for potatoes, a household that did not ordinarily barter might do so. Such considerations may have been more important under the economy of shortage and may be of lesser significance today—except perhaps during the hyperinflation of 1997.

9. A less conspiratorial explanation is also possible. The traders might have had as their objective the sale of goods, rather than the procurement of potatoes, and thus they may have simply been accepting potatoes for their goods rather than for money, due to the villagers' lack of cash.

10. I later learned that some of these village traders are typically among those classified as rich by other villagers.

11. One weaver reported that during the crisis's peak she was reluctant to sell the rugs and blankets that she made even for hard currency. She felt that it was better to have the products, because they would hold their value better than money.

12. Accompanying a friend to the milk-buying point one day to pick up the household's milk money, we met a woman who was collecting money for water bills.

13. The cherries had come from the trees of a family friend who lived in the lowlands.

14. During the socialist era, celebration of religious holidays was discouraged. In the case of Christmas, the gift giving was moved to New Year's Day, and the gifts were delivered by Grandfather Frost rather than by Grandfather Christmas.

15. My landlady, for example, frequently observed that there has not been a "real" wedding in the village in ten years, the implication here being that people cannot afford the expense of a big party to which everyone is invited. Similarly, many villagers were able to build and furnish new houses during the socialist era, while many houses begun since then sit as empty, half-constructed shells on the hillside while their owners look for buyers because they do not have the resources to complete construction.

16. In some sense, this focus on financial stratification and nostalgia for when they were all equal ignores the fact that socialist-era stratification may have been based on different criteria, such as the number and nature of one's connections.

17. I did not specifically explore these issues during 1997 because the economy was in such disarray at the time.

18. One person said that being "average" in the village means being poor in the larger picture, but then clarified this by saying that village life did have certain advantages over that in town. In particular, village residents produce some of their own food so that they have meat and dairy products, which might be unaffordable to some town dwellers, and they also can avoid paying for central heating because they have ready access to firewood.

5. Conserving the Natural Heritage of the Rhodope

1. In distinguishing between rural and urban, I do not mean to imply that no middle ground exists between the two. Yet, the division does have analytical and practical significance in the Bulgarian context (see also Creed 1993 on this point).

2. Plovdiv, Bulgaria's second-largest city, is located about two hours east of Sofia by bus, train, or car, and the cities are connected by highway.

3. This was again an issue in summer 2001, and both sides of the issue received coverage on national TV.

4. At issue was the change in status from "national park" to "nature park."

5. Village residents cited membership in two organizations: the local commercial cooperative and the local hunting and fishing society. Membership in the latter group is required for people wanting to hunt or fish.

6. Bridger (1998:207–208) gives a similar example where an individual's experience on a small-business study tour to the United States was subsequently incorporated into the activities of a women's organization in Russia.

7. At least two other kinds of connections are possible, although they are less critical to an understanding of the links between the global and

the local and thus are discussed here only briefly. Bulgarian environmental organizations at the same level sometimes cooperate among
themselves. Due in part to competition over funds and perhaps a lack of
experience with cooperation, this has taken place more over immediate
and pressing environmental threats or in response to funding explicitly
designated for cooperative projects and less frequently has occurred on
a long-term or institutionalized basis. More recently, cooperative efforts
among some groups seem to be more frequent. For example, the Ministry of the Environment awarded a project on the conservation of the
imperial eagle to a coalition of NGOs and other partners including
Green Balkans, the Bulgarian Society for the Protection of Birds, the
Bulgarian Bird Ringing Center, the Wilderness Fund, the biology faculty
at Sofia University, and the Plovdiv Museum of Natural History. As a
leader in one group put it, they are now experienced enough to recognize that they have worse enemies than each other. The second form of
cooperation is between organizations at approximately the same level in
neighboring countries, sometimes again in response to funding opportunities for cooperative projects, but also in recognition of the fact that
mountain ranges and animal populations are not necessarily constrained by political borders.

8. The World Wide Fund for Nature, established in 1961 as the World
Wildlife Fund, is headquartered in Gland, Switzerland, and has national
organizations in many countries worldwide. The WWF usually works
through NGOs based in individual countries and also through country-
based program offices. The national affiliates in the United States and
Canada continue to use the name World Wildlife Fund.

9. The UK-based Royal Society for the Protection of Birds is a membership-based organization that was established in the 1890s. Although
primarily a national organization, RSPB also has an international program and participates in BirdLife International, a global partnership of
60 bird and habitat protection organizations worldwide. As such it partners with a certain number of lesser-developed countries, including
Bulgaria, providing organizations in these countries with financial and
technical support along with training. The RSPB is also responsible for
the training activities of the entire BirdLife partnership.

10. The World Conservation Union or IUCN, founded in 1948, is a
union of state, government agency, and NGO members along with a
global network of environmental expertise composed of IUCN staff, volunteer scientists, and practitioners. Headquartered in Gland, Switzerland, the organization has regional and country offices scattered around
the world. The IUCN was perhaps the only international conservation
organization with a formal presence in Soviet-bloc countries during the
socialist era, a fact often attributed to its identification more as a scien-

tific or technical organization rather than as an environmental advocacy group (IUCN 1996/97).

11. Raising local varieties of livestock has also been promoted by some of the NGOs, for example, Karakchanov sheep and goats. These animals, which are well adapted to local conditions, may have higher market values, and by preserving the varieties, which went out of style during collectivization, local biodiversity is enhanced.

12. Similar efforts have promoted the cultivation of higher-value alternatives to the more commonly produced potatoes and tobacco, such as berries or organic produce. To my knowledge, cultivation of the wild mushroom varieties is not considered feasible. But in one case, an NGO has also encouraged community groups to purchase, process, and market wild-collected products so that they are able to capture some of the profit margin otherwise going to middlemen.

6. A Civil Balkan Village? Cavers and Collective Action

1. Some examples from the small library at the research institute where this manuscript was developed include Bermeo and Nord 2000 (*Civil Society before Democracy: Lessons from Nineteenth-Century Europe*); Brook and Frolic 1997 (*Civil Society in China*); Comaroff and Comaroff 1999 (*Civil Society and the Political Imagination in Africa*); Ekiert and Kubik 1999 (*Rebellious Civil Society: Popular Protest and Democratic Consolidation in Poland, 1989–1993*); Gellner 1994 (*Conditions of Liberty: Civil Society and Its Rivals*); Hall 1996 (*Civil Society: Theory, History, Comparison*); Hann and Dunn 1996 (*Civil Society: Challenging Western Models*); Keane 1998 (*Civil Society and the State: New European Perspectives*); Monga 1996 (*The Anthropology of Anger: Civil Society and Democracy in Africa*); Ndegwa 1996 (*The Two Faces of Civil Society: NGOs and Politics in Africa*); Ruffin 1999 (*Civil Society in Central Asia*); and Tismaneanu 1995 (*Political Culture and Civil Society in Russia and the New States of Eurasia*). A bigger library or search of journal articles would produce many more examples.

2. Indeed, one reason that I did not pick the village as the site for my extended fieldwork was concern that the road would wash out, periodically isolating the village or at least making traveling more difficult. Such concern was not necessarily unwarranted. The two most frequent topics to emerge from Internet searches for the village name in Bulgarian newspapers are the NGO and tourism activities discussed in this chapter, and road closures and what is being done to remedy them.

3. Presumably one connection here is that such individuals had technical experience relevant to club activities.

4. This association unites more than a dozen village households offering tourist accommodations. Its primary activities appear to consist

of attending tourism seminars sponsored or organized by other Bulgarian NGOs and hosting tourists as a private business activity of individual member households. I first met its president on a visit to the village
with officials from the Bulgarian Association for Rural and Ecological
Tourism before the local group had been formed, and the local association identifies this Sofia-based NGO as an important resource.

5. Most of the international expeditions took place before the collapse of communism in Central and Eastern Europe and the Soviet
Union, hence my use of the term "socialist-bloc countries" without the
qualifier "former." During the socialist era, travel within the socialist-
bloc was often easier than travel to the capitalist West. In the postsocialist period economic constraints make international travel difficult.

6. According to the club president, the receipts from this site go to the
municipality and the club itself does not receive any of the money.

7. In part this may stem for the death of the society's founding president in 1997 and its recent difficulties with mobilizing members. The decline in activities is not entirely the result of the loss of a person who was
a driving force behind the organization. For example, 1997 was a year
of acute economic crisis in Bulgaria, and the chronically poor economic
situation in Bulgaria during the 1990s has undoubtedly diverted the
energy of some people. Yet the absence of this charismatic and energetic
leader is undoubtedly an important factor. The apparent fate of the Trigrad branch is not necessarily unique, however. Following his death, the
leaders in Borovo similarly expressed concern about the future of their
branch.

8. A recent paper makes a similar argument about the role of what its
author calls "trail-based companionship" in early conservation efforts
and organizations in the United States (Greenfield 2001).

For my most recent Bulgarian research I have been spending
time in Chepelare, a central Rhodope town that is the municipal center
with which Zaburdo is connected and which has about 6,000 residents.
Like Trigrad, this town had and still has a socialist-era caving club, albeit
with the usual postsocialist financial difficulties. Without any intention
of doing so, I keep running into Chepelare residents who relate fond
memories of their experiences in the club in their youth—about the expeditions in which they participated and sometimes also of the scientific
investigations conducted on these expeditions. For them it clearly was
an important community activity, not just some socialist nonsense.

9. An example is the current president of the caving federation, who
was active in the organization Ecoglasnost at the time of the political
changes.

10. Here we are reminded of the increased educational opportunities
available in the socialist era including those beyond the high-school

level. With the "industrialization" of agriculture, there were also job opportunities for at least some of the graduates in rural settings.

11. Compare here my description of the caving club with the leadership of the BSCRM branches in the predominantly Christian villages of Borovo or Shiroka Luka, for example.

Bibliography

State Gazette

Following passage, new Bulgarian laws are published in the *State Gazette* (*Durzhaven Vestnik*)—essentially equivalent to the *Federal Register* in the United States. I have provided these references as in-text citations for interested readers; they consist of the date of publication along with the issue within the given year of the *State Gazette*.

Archival and Unpublished Materials

In the preparation of this work, I occasionally drew upon unpublished materials, both from archives and from various government offices. I explain here briefly the nature of these sources.

All cited archival materials are held in the Smolyan Regional Archives, stock number 372K, 373K, and 894. These archives are located in the town of Smolyan; however researchers may need to obtain permission from the central archives in Sofia before they can be used. (That was the case the last time I used them).

Other unpublished materials came from a variety of government offices including the village administration in Zaburdo, the National Statistical Institute in Sofia, the National Institute of Meteorology and Hydrology at the Bulgarian Academy of Sciences in Sofia, and the Chepelare Municipal Land Commission. Because these materials are unpublished, they lack publication dates. When possible, however, I have indicated the year in which I received the data.

Books and Articles

Abramson, David M. 1999. A Critical Look at NGOs and Civil Society as Means to an End in Uzbekistan. *Human Organization* 58(3):240–250.

Agrawal, Arun, and Clark C. Gibson, eds. 2001. *Communities and the Environment: Ethnicity, Gender, and the State in Community-Based Conservation.* New Brunswick NJ: Rutgers University Press.

Allin, Craig W. 1990. Introduction: National Parks and Nature Reserves in Global Perspective. In *International Handbook of National Parks and Nature Reserves.* Craig W. Allin, ed. Pp. 1–20. New York: Greenwood Press.

Amadeus, zashtiti Vitosha! (Amadeus, protect Vitosha!). 1997. Editorial. *24 Chasa,* December 18:10.

Appadurai, Arjun. 1986. Introduction: Commodities and the Politics of Value. In *The Social Life of Things: Commodities in Cultural Perspective.* Arjun Appadurai, ed. Pp. 3–63. Cambridge: Cambridge University Press.

Arellano-López, Sonia, and James F. Petras. 1994. Non-Governmental Organizations and Poverty Alleviation in Bolivia. *Development and Change* 25:555–568.

Baker, Susan, and Bernd Baumgartl. 1998. Bulgaria: Managing the Environment in an Unstable Transition. *Environmental Politics* 7(1):183–206.

Barnes, Robert H., and Ruth Barnes. 1989. Barter and Money in an Indonesian Village Economy. *Man* 24:399–418.

Bates, Daniel G. 1995. Uneasy Accommodation: Ethnicity and Politics in Rural Bulgaria. In *East European Communities: The Struggle for Balance in Turbulent Times.* David A. Kideckel, ed. Pp. 137–157. Boulder CO: Westview Press.

Baumgartl, Bernd. 1993. Environmental Protest as a Vehicle for Transition: The Case of EkoGlasnost in Bulgaria. In *Environment and Democratic Transition: Policy and Politics in Central and Eastern Europe.* Anna Vari and Pal Tamas, eds. Pp. 157–175. Dordrecht: Kluwer Academic Publishers.

Beck, Sam. 1976. The Emergence of the Peasant Worker in a Transylvanian Mountain Community. *Dialectical Anthropology* 1(4):365–375.

Begg, Robert, and Mieke Meurs. 1998. Writing a New Song: Path Dependency and State Policy in Reforming Bulgarian Agriculture. In *Privatizing the Land: Rural Political Economy in Post-Communist Societies.* Iván Szelényi, ed. Pp. 245–270. London: Routledge.

Bell, John D. 1977. *Peasants in Power: Alexander Stamboliski and the Bulgarian Agrarian National Union, 1988–1923.* Princeton NJ: Princeton University Press.

———. 1998. Appendix. In *Bulgaria in Transition: Politics, Economics, So-*

ciety, and Culture after Communism. John D. Bell, ed. Pp. 327–332. Boulder CO: Westview Press.

Benthall, Jonathan. 2000. Civil Society's Need for De-Deconstruction. *Anthropology Today* 16(2):1–4.

Berdahl, Daphne, Matti Bunzl, and Martha Lampland, eds. 2000. *Altering States: Ethnographies of Transition in Eastern Europe and the Former Soviet Union.* Ann Arbor: University of Michigan Press.

Berglund, Eeva, and Krista Harper. 2001. Citizen Sensibilities: Comparing Grassroots Environmental Activism in Finland and Hungary. *Anthropology of East Europe Review* 19(1):5–10.

Bermeo, Nancy Gina, and Philip G. Nord, eds. 2000. *Civil Society before Democracy: Lessons from Nineteenth-Century Europe.* Lanham MD: Rowman and Littlefield.

Bishop, S., K. Howe, D. Kopeva, and P. Mishev. 1994. Land Markets and Tenure. In *Privatization of Agriculture in New Market Economies: Lessons from Bulgaria.* Andrew Schmitz, Kirby Moulton, Allan Buckwell, and Sofia Davidova, eds. Pp. 75–86. Boston: Kluwer Academic Publishers.

Blaikie, Piers. 1995. Changing Environments or Changing Views? A Political Ecology for Developing Countries. *Geography* 80(3):203–214.

Blaikie, Piers, and Harold Brookfield. 1987. *Land Degradation and Society.* London: Methuen.

Bogdanova, Iskra. 1977. C liubov i grizha (With love and concern). *Narodna Mladezh,* February 20.

Bohannan, Paul. 1955. Some Principles of Exchange and Investment among the Tiv. *American Anthropologist* 57:60–70.

Bojinov, Christo, Alexander Alexandrov, Zdravko Vassiliev, Georgy Raffailov, Boyan Rossnev, and Meglena Plougchieva. 1994. *Forest and Forest Products Country Profile: Bulgaria.* New York: United Nations.

Bonner, Raymond. 1993. *At the Hand of Man: Peril and Hope for Africa's Wildlife.* New York: Alfred A. Knopf.

Bridger, Sue. 1998. Tackling the Market: The Experience of Three Moscow Women's Organizations. In *Surviving Post-Socialism: Local Strategies and Regional Responses.* Sue Bridger and Frances Pine, eds. Pp. 203–218. London: Routledge.

Bridger, Sue, and Frances Pine, eds. 1998. *Surviving Post-Socialism: Local Strategies and Regional Responses.* London: Routledge.

Brook, Timothy, and B. Michael Frolic, eds. 1997. *Civil Society in China.* Armonk NY: Sharpe.

Brower, Barbara Anne. 1991. *Sherpa of Khumbu: People, Livestock and Landscape.* Delhi: Oxford University Press.

Brown, Jessica, and Brent Mitchell. 1994. *Landscape Conservation and Stewardship in Central Europe.* Nexus Occasional Paper, No. 10. Ipswich MA: Atlantic Center for the Environment.

Brown, Michael, and Barbara Wyckoff-Baird. 1992. *Designing Integrated Conservation and Development Projects*. Washington DC: Biodiversity Support Program.

Brunnbauer, Ulf. 1998. Histories and Identities: Nation–State and Minority Discourses: The Case of the Bulgarian Pomaks. In *In and Out of the Collective: Papers on Former Soviet Bloc Rural Communities*, vol. 1. Electronic document, http://www.nbu.bg/iafr/ulf.htm, accessed March 25, 2004.

Bruno, Marta. 1998. Playing the Co-Operation Game: Strategies around International Aid in Post-Socialist Russia. In *Surviving Post-Socialism: Local Strategies and Regional Responses*. Sue Bridger and Frances Pine, eds. Pp. 170–187. London: Routledge.

Buchanan, Donna Anne. 1991. The Bulgarian Folk Orchestra: Cultural Performance, Symbol, and the Construction of National Identity in Socialist Bulgaria. Ph.D. dissertation, University of Texas, Austin.

Bulgarian NGOs: Towards the NGO Parallel Conference "Environment for Europe." 1995. Unpublished mimeographed document, Borrowed Nature Association, Sofia.

Bulgarian Society for the Conservation of the Rhodope Mountains and World Wide Fund for Nature. 1995. *Planning for Conservation: Participatory Rural Appraisal for Community Based Initiatives*. Report on the PRA Training Workshop held in Ostritza, Bulgaria, 14–22 June 1993. Gland: WWF-International.

Bulgarian Society for the Protection of Birds. 1996. Protecting the National Park PIRIN—A World Heritage Site. *Neophron: Information Bulletin of the Bulgarian Society for the Protection of Birds* (English-language version) 1:3.

Burawoy, Michael, and Katherine Verdery. 1999. Introduction. In *Uncertain Transition: Ethnographies of Change in the Postsocialist World*. Michael Burawoy and Katherine Verdery, eds. Pp. 1–17. Lanham MD: Rowman and Littlefield.

Carroll, Thomas F. 1992. *Intermediary NGOs: The Supporting Link in Grassroots Development*. West Hartford CT: Kumarian Press.

Carter, F. W., and D. Turnock, eds. 1993. *Environmental Problems in Eastern Europe*. New York: Routledge.

Caryl, Christian. 1998. Siberian Macaroni: Pasta King Explains How Russia's Economy Really Works. *U.S. News and World Report* 124(24):42–44.

Cellarius, Barbara A. 1998. Linking Global Priorities and Local Realities: Nongovernmental Organizations and the Conservation of Nature in Bulgaria. In *Bulgaria in Transition: Environmental Consequences of Political and Economic Transformation*. Krassimira Paskaleva, Philip Shapira,

John Pickles, and Boian Koulov, eds. Pp. 57–82. Aldershot: Ashgate Press.

———. 2001. Seeing the Forest for the Trees: Local-Level Resource Use and Forest Restitution in Postsocialist Bulgaria. *GeoJournal.* 55(2–4): 599–606.

Cellarius, Barbara A., and Caedmon Staddon. 2002. Environmental Nongovernmental Organizations, Civil Society and Democratization in Bulgaria. *East European Politics and Societies* 16(1):182–222.

Chapman, Ann. 1980. Barter as a Universal Mode of Exchange. *L'Homme* 20(3):33–83.

Clark, Mari H. 1997. Wild Herbs in the Marketplace: Gathering in Response to Market Demand. In *Aegean Strategies: Studies of Culture and Environment on the European Fringe.* P. Nick Kardulias and Mark T. Shutes, eds. Pp. 215–236. Lanham MD: Rowman and Littlefield.

Close, David H. 1998. Environmental NGOs in Greece: The Acheelöos Campaign as a Case Study of Their Influence. *Environmental Politics* 7(2):55–77.

———. 1999. Environmental Movements and the Emergence of Civil Society in Greece. *Australian Journal of Politics and History* 45(1):52–64.

Cole, John W. 1977. Anthropology Comes Part-Way Home—Community Studies in Europe. *Annual Review of Anthropology* 6:349–378.

Cole, John W., and Eric R. Wolf. 1974. *The Hidden Frontier: Ecology and Ethnicity in an Alpine Valley.* New York: Academic Press.

Colson, Elizabeth, and Conrad P. Kottak. 1996. Linkages Methodology for the Study of Sociocultural Transformations. In *Transforming Societies, Transforming Anthropology.* Emilio F. Moran, ed. Pp. 103–134. Ann Arbor: University of Michigan Press.

Comaroff, John L., and Jean Comaroff, eds. 1999. *Civil Society and the Political Imagination in Africa: Critical Perspectives.* Chicago: University of Chicago Press.

Crampton, Richard J. 1987. *A Short History of Modern Bulgaria.* Cambridge: Cambridge University Press.

———. 1990. The Intelligentsia, the Ecology, and the Opposition in Bulgaria. *World Today* 46(2):23–26.

———. 1997. *A Concise History of Bulgaria.* Cambridge: Cambridge University Press.

Creed, Gerald W. 1991. Between Economy and Ideology: Local-Level Perspectives on Political and Economic Reform in Bulgaria. *Socialism and Democracy* 13:45–65.

———. 1993. Rural–Urban Oppositions in the Bulgarian Political Transition. *Südosteuropa* 42(6):369–382.

———. 1995a. Agriculture and the Domestication of Industry in Rural Bulgaria. *American Ethnologist* 22(3):528–548.

————. 1995b. An Old Song in a New Voice: Decollectivization in Bulgaria. In *East European Communities: The Struggle for Balance in Turbulent Times*. David A. Kideckel, ed. Pp. 25–45. Boulder CO: Westview Press.

————. 1995c. The Politics of Agriculture: Identity and Socialist Sentiment in Bulgaria. *Slavic Review* 54:843–68.

————. 1998. *Domesticating Revolution: From Socialist Reform to Ambivalent Transition in a Bulgarian Village*. University Park: Pennsylvania State University Press.

Creed, Gerald W., and Janine R. Wedel. 1997. Second Thoughts from the Second World: Interpreting Aid in Post-Communist Eastern Europe. *Human Organization* 56(3):253–264.

Danchev, Jordan. 1998. Report of the Bulgarian Union for the Conservation of the Rhodope Mountains. In *Bulgaria's Biological Diversity: Conservation Status and Needs Assessment*, vols. 1 and 2. Curt Meine, ed. Pp. 741–750. Washington DC: Biodiversity Support Program.

Daskalov, Roumen. 1998. A Democracy Born in Pain: Bulgarian Politics, 1989–1997. In *Bulgaria in Transition: Politics, Economics, Society and Culture after Communism*. John D. Bell, ed. Pp. 9–38. Boulder CO: Westview Press.

Dechov, Vasil. 1903. Proizvodstvo na len v srednite Rodopi (Production of linen in the middle Rhodope). *Rodopski Napreduk* 1(1):273–275.

————. 1968. *Izbrani suchineniya* (Selected works). Plovdiv: Izdatelstvo Hristo G. Danov.

Desai, Uday, and Keith Snavely. 1998. Emergence and Development of Bulgaria's Environmental Movement. *Nonprofit and Voluntary Sector Quarterly* 27(1):32–48.

De Soto, Hermine G., and Nora Dudwick, eds. 2000. *Fieldwork Dilemmas: Anthropologists in Postsocialist States*. Madison: University of Wisconsin Press.

Dimitrov, Valentin. 1976. Mladite v zashtita na prirodata (Youths in the protection of nature). *Narodna Mladezh*, December 24.

Doane, Molly. 2001. A Distant Jaguar: The Civil Society Project in Chimalapas. *Critique of Anthropology* 21(4):361–382.

Dobreva, Stanka. 1994. The Family Farm in Bulgaria: Traditions and Changes. *Sociologia Ruralis* 34(4):340–353.

Donchev, Anton. 1997. *Vreme Razdelno* (A Time of Division). Sofia: Zaharii Stoyanov.

Drabek, Anne Gordon, ed. 1987. Special supplement on NGOs and Development. *World Development* 15.

Dunn, Elizabeth C. 1998. "A Product for Everybody is a Product for Nobody": Niche Marketing and Political Individualism in Polish Civil Society. *Anthropology Today* 14(4):22–23.

Ecological Bricks for Our Common House of Europe. 1990. *Politische Ökologie* 2.

The Economist. 1997. The Cashless Society. *The Economist,* March 15:77–78.

Edwards, Michael, and David Hulme, eds. 1992. *Making a Difference: NGOs and Development in a Changing World.* London: Earthscan.

———. 1996. *Beyond the Magic Bullet: NGO Performance and Accountability in the Post–Cold War World.* West Hartford CT: Kumarian Press.

Ekiert, Grzegorz, and Jan Kubik. 1999. *Rebellious Civil Society: Popular Protest and Democratic Consolidation in Poland, 1989–1993.* Ann Arbor: University of Michigan Press.

Ellis, Frank. 1998. Household Strategies and Rural Livelihood Diversification. *Journal of Development Studies* 35(1):1–38.

Eminov, Ali. 1997. *Turkish and Other Muslim Minorities in Bulgaria.* New York: Routledge.

Feeny, David, Fikret Berkes, Bonnie J. McCay, and James M. Acheson. 1990. The Tragedy of the Commons: Twenty-Two Years Later. *Human Ecology* 18(1):1–19.

Fisher, Duncan. 1993. The Emergence of the Environmental Movement in Eastern Europe and Its Role in the Revolutions of 1989. In *Environmental Action in Eastern Europe: Responses to Crisis.* Barbara Jancar-Webster, ed. Pp. 89–113. Armonk NY: M. E. Sharpe.

Fisher, Julie. 1996. Grassroots Organizations and Grassroots Support Organizations: Patterns of Interaction. In *Transforming Societies, Transforming Anthropology.* Emilio F. Moran, ed. Pp. 57–102. Ann Arbor: University of Michigan Press.

Fisher, William F. 1997. Doing Good? The Politics and Antipolitics of NGO Practices. *Annual Review of Anthropology* 26:439–464.

Flanz, Gisbert H. 1992. Republic of Bulgaria. In *Constitutions of the Countries of the World.* Albert P. Blaustein and Gisbert H. Flanz, eds. Pp. 87–120. Dobbs Ferry NY: Oceana Publications.

Forbes, Hamish. 1997. A "Waste" of Resources: Aspects of Landscape Exploitation in Lowland Greek Agriculture. In *Aegean Strategies: Studies of Culture and Environment on the European Fringe.* P. Nick Kardulias and Mark T. Shutes, eds. Pp. 187–214. Lanham MD: Rowman and Littlefield.

French, Hillary F. 1995. Forging a New Global Partnership. In *State of the World 5: A Worldwatch Institute Report on Progress Toward a Sustainable Society.* Linda Starke, ed. Pp. 170–189. New York: W. W. Norton.

Friedberg, James, and Branimir Zaimov. 1998. Politics, Environment and the Rule of Law in Bulgaria. In *Bulgaria in Transition: Environmental Consequences of Political and Economic Transformation.* Krassimira Paskaleva, Philip Shapira, John Pickles, and Boian Koulov, eds. Pp. 83–103. Aldershot: Ashgate.

Fry, William E., and Stephen B. Goodwin. 1997. Resurgence of the Irish Potato Famine Fungus. *BioScience* 47(6):363–371.

GEF Secretariat. 1999. Introduction to the Global Environmental Facility. Electronic document, http://www.gefweb.org/intro/gefintro.htm, accessed June 1999.

Gellner, Ernest. 1994. *Conditions of Liberty: Civil Society and Its Rivals*. New York: Allen Lane/Penguin Press.

Genov, Nikolai. 1993. Environmental Risks in a Society in Transition: Perceptions and Reactions. In *Environment and Democratic Transition: Policy and Politics in Central and Eastern Europe*. Anna Vari and Pal Tamas, eds. Pp. 268–280. Dordrecht: Kluwer Academic Publishers.

Georgiev, Georgi. 1993. *Narodnite parkove i rezervatite v Bulgariya* (The national parks and reserves in Bulgaria). Sofia: Prosveta.

Georgiev, Velichko, and Stayko Trifonov. 1995. *Pokrustvaneto na Bulgarite Mohamedani, 1912–1913, Dokumenti* (The conversion of the Bulgarian Muslims, 1912–1913, Documents). Sofia: Akademichno Izdatelstvo Prof. Marin Drinov.

Georgieva, Kristalina. 1993. Environmental Policy in a Transition Economy: The Bulgarian Economy. In *Environment and Democratic Transition: Policy and Politics in Central and Eastern Europe*. Anna Vari and Pal Tamas, eds. Pp. 67–87. Dordrecht: Kluwer Academic Publishers.

Georgieva, Kristalina, and Judith Moore. 1997. Bulgaria. In *The Environmental Challenge for Central European Economies in Transition*. Jürg Klarer and Bedrich Moldan, eds. Pp. 67–106. New York: John Wiley and Sons.

Gezon, Lisa L. 1997. Institutional Structure and the Effectiveness of Integrated Conservation and Development Projects: Case Study from Madagascar. *Human Organization* 56(4):462–470.

Ghimire, Krishna B. 1994. Parks and People: Livelihood Issues in National Parks Management in Thailand and Madagascar. *Development and Change* 25:195–229.

Giordano, Christian. 1993. Not All Roads Lead to Rome. *Eastern European Countryside* 1(1):5–16.

Goldman, Marshall I. 1998. The Cashless Society. *Current History* 97(621):319–324.

Grant, Bruce. 1995. *In the Soviet House of Culture: A Century of Perestroikas*. Princeton NJ: Princeton University Press.

Greenberg, James B., and Thomas K. Park. 1994. Political Ecology. *Journal of Political Ecology* 1:1–10.

Greenfield, Joshua. 2001. "Good Fellowship and Camaraderie": Group Hiking Trips and the Growth of the Early American Conservation Movement. Unpublished master's thesis, Hunter College, City University of New York.

Green Patrols. N.d. Declaration. Unpublished mimeographed document.

Grimm, Curt, and Bruce Byers. 1994. NGOs and the Integration of Conservation and Development in Madagascar: An Assessment for the USAID SACVEM Project. *Development Anthropology Network* 12(1/2):30–38.

Grodeland, Ase B., Tatyana Y. Koshechkina, and William L. Miller. 1998. "Foolish to Give and Yet More Foolish Not to Take"—In-Depth Interviews with Post-Communist Citizens on Their Everyday Use of Bribes and Contacts. *Europe-Asia Studies* 50(4):651–677.

Guha, Ramachandra. 1997. The Environmentalism of the Poor. In *Between Resistance and Revolution: Cultural Politics and Social Protest*. Richard G. Fox and Orin Starn, eds. Pp. 17–39. New Brunswick NJ: Rutgers University Press.

Gumnerova, Mariya. 1969. Tukani ot len i konop v Smolyansko (Textiles from flax and hemp in the Smolyan area). In *Rodopski Sbornik*, vol. 2. Hr. Hristov, P. Petrov, and Str. Dimitrov, eds. Pp. 131–145. Sofia: Izdatelstvo na Bulgarskata Akademiya na Naukite.

Hadjiev, I. T. 1969. Sanitarno-higienna anketa na c. Zaburdo, Smolyanski okrug (Sanitary-hygiene study of the village of Zaburdo, Smolyan region). Unpublished thesis, Vissh Meditsinski Institut I. P. Pavlov.

Haitov, Nikolai. 1979. *Wild Tales*. Michael Holman, trans. London: Peter Owen.

Hall, John A., ed. 1996. *Civil Society: Theory, History, Comparison*. Cambridge: Polity Press.

Halpern, Joel Martin, and David A. Kideckel. 1983. Anthropology of Eastern Europe. *Annual Review of Anthropology* 12:377–402.

Hann, C. M. 1980. *Tázlár: A Village in Hungary*. Cambridge: Cambridge University Press.

———. 1987. The Politics of Anthropology in Socialist Eastern Europe. In *Anthropology at Home*. Anthony Jackson, ed. Pp. 139–153 London: Tavistock Publications.

———. 1993a. From Comrades to Lawyers: Continuity and Change in Local Political Culture in Rural Hungary. *Anthropological Journal of European Cultures* 2(1):75–104.

———. 1993b. From Production to Property: Decollectivization and the Family–Land Relationship in Contemporary Hungary. *Man*, n.s. 28:299–320.

———. 1993c. Introduction: Social Anthropology and Socialism. In *Socialism: Ideals, Ideologies, and Local Practice*. C. M. Hann, ed. Pp. 1–26. London: Routledge.

———. 1994. After Communism: Reflections on East European Anthropology and the "Transition." *Social Anthropology* 2(3):229–249.

Hann, C. M., ed. 2002. *Postsocialism: Ideals, Ideologies, and Practices in Eurasia*. London: Routledge.

Hann, C. M., and Elizabeth Dunn, eds. 1996. *Civil Society: Challenging Western Models*. London: Routledge.

Hardalova, Rayna, Lyuba Evstatieva, and Chavdar Gussev. 1998. Wild Medicinal Plant Resources in Bulgaria and Recommendations for Their Long-Term Development. In *Bulgaria's Biological Diversity: Conservation Status and Needs Assessment*, vols. 1 and 2. Curt Meine, ed. Pp. 527–561. Washington DC: Biodiversity Support Program.

Hardin, Garrett. 1968. The Tragedy of the Commons. *Science* 162:1243–1248.

Heinen, Joel T., and Bijaya Kattel. 1992. Parks, People, and Conservation: A Review of Management Issues in Nepal's Protected Areas. *Population and Environment* 14(1):49–84.

Hemment, Julie. 1998. Colonization or Liberation: The Paradox of NGOs in Postsocialist States. *Anthropology of East Europe Review* 16(1):31–39.

———. 2000. The Price of Partnership: The NGO, the State, the Foundation, and Its Lovers in Post-Communist Russia. *Anthropology of East Europe Review* 18(1):33–36.

Hitchcock, Robert K. 1995. Centralization, Resource Depletion, and Coercive Conservation among the Tyua of the Northeastern Kalahari. *Human Ecology* 23(2):169–198.

Ho, Peter. 2001. Greening without Conflict? Environmentalism, NGOs, and Civil Society in China. *Development and Change* 32(5):893–921.

Homewood, K. M., and W. A. Rodgers. 1991. *Maasailand Ecology: Pastoralist Development and Wildlife Conservation in Ngorongoro, Tanzania*. Cambridge: Cambridge University Press.

Hulme, David, and Marshall Murphree, eds. 2001. *African Wildlife and Livelihoods: The Promise and Performance of Community Conservation*. Oxford: James Currey Publishers.

Humphrey, Caroline. 1985. Barter and Economic Disintegration. *Man* 20:48–72.

———. 1992. Fair Dealing, Just Rewards: The Ethics of Barter in North-East Nepal. In *Barter, Exchange, and Value: An Anthropological Approach*. Caroline Humphrey and Stephen Hugh-Jones, eds. Pp. 107–141. Cambridge: Cambridge University Press.

———. 1998 [1983]. *Marx Went Away — But Karl Stayed Behind* (Updated edition of *Karl Marx Collective: Economy, Society and Religion in a Siberian Collective Farm*). Ann Arbor: University of Michigan Press.

———. 1999. Traders, "Disorder," and Citizenship Regimes in Provincial Russia. In *Uncertain Transition: Ethnographies of Change in the Postsocialist World*. Michael Burawoy and Katherine Verdery, eds. Pp. 19–52. Lanham MD: Rowman and Littlefield.

————. 2000. An Anthropological View of Barter in Russia: In *The Van-ishing Rouble: Barter Networks and Non-Monetary Transactions in Post-Soviet Societies*. Paul Seabright, ed. Pp. 71–90. Cambridge: Cambridge University Press.

Humphrey, Caroline, and Stephen Hugh-Jones. 1992. Introduction: Barter, Exchange, and Value. In *Barter, Exchange, and Value: An Anthropological Approach*. Caroline Humphrey and Stephen Hugh-Jones, eds. Pp. 1–20. Cambridge: Cambridge University Press.

Hussein, Karim, and John Nelson. 1998. Sustainable Livelihoods and Livelihood Diversification. IDS Working Paper 69. Brighton: Institute of Development Studies.

Iliev, Chavdar. 1973. *Selo Zaburdo: Istoriya, bit, narodno tvorchestvo* (The village of Zaburdo: History, customs, and folklore). Sofia: Izdatelstvo na otechestveniya front.

————. 1981. Spomeni za osnovavaneto i deynostta na narodno chital-ishte "Hristo Botev" i novi danni za otkrivane na uchilishteto v c. Zaburdo (Recollections about the founding and the activities of the cultural center "Hristo Botev" and new information about the opening of the school in the village of Zaburdo). Unpublished MS.

Iordanova, Dina. 1998. Canaries and Birds of Prey: The New Season of Bulgarian Cinema. In *Bulgaria in Transition: Politics, Economics, Society, and Culture after Communism*. John D. Bell, ed. Pp. 255–277. Boulder CO: Westview Press.

Islamoglu, Huri. 2000. Property as Contested Domain: A Reevaluation of the Ottoman Land Code of 1858. In *New Perspectives on Property and Land in the Middle East*. Roger Owen, ed. Pp. 3–61. Cambridge MA: Harvard University Press.

IUCN (World Conservation Union). 1991. *Protected Areas of the World: A Review of National Systems*, vol. 2: *Palaearctic*. Gland: IUCN.

————. 1994a. *1993 United Nations List of National Parks and Protected Areas*. Gland: IUCN.

————. 1994b. *Parks for Life: Action for Protected Areas in Europe*. Gland: IUCN.

————. 1995. *The Mountains of Central and Eastern Europe*. Gland: IUCN.

————. 1996/1997. *A Pocket Guide to IUCN: The World Conservation Union*. Gland: IUCN.

————. 1998. *1997 United Nations List of Protected Areas*. Gland: IUCN.

IUCN/UNEP/WWF (World Conservation Union/United Nations Environmental Program/World Wide Fund for Nature). 1991. *Caring for the Earth: A Strategy for Sustainable Living*. Gland: IUCN.

Jancar-Webster, Barbara. 1993. The East European Environmental Movement and the Transformation of East European Society. In *Environmental Action in Eastern Europe: Responses to Crisis*. Barbara Jancar-Webster, ed. Pp. 192–219. Armonk NY: M. E. Sharpe.

————. 1998. Environmental Movement and Social Change in the Transition Countries. *Environmental Politics* 7(1):69–90.

Jelavich, Barbara. 1983. *History of the Balkans,* vol. 2: *Twentieth Century.* Cambridge: Cambridge University Press.

Jorgens, Denise. 2000. A Comparative Examination of the Provisions of the Ottoman Land Code and Khedive Sa'id's Law of 1858. In *New Perspectives on Property and Land in the Middle East.* Roger Owen, ed. Pp. 93–119. Cambridge MA: Harvard University Press.

Junghans, Trenholme. 2001. Marketing Selves: Constructing Civil Society and Selfhood in Post-Socialist Hungary. *Critique of Anthropology* 21(4):383–400.

Kaneff, Deema. 1996. Responses to "Democratic" Land Reforms in a Bulgarian Village. In *After Socialism: Land Reform and Social Change in Eastern Europe.* Ray Abrahams, ed. Pp. 85–114. Oxford: Berghahn Books.

————. 1998. When "Land" Becomes "Territory": Land Privatization and Ethnicity in Rural Bulgaria. In *Surviving Post-Socialism: Local Strategies and Regional Responses.* Frances Pine and Sue Bridger, eds. Pp. 16–32. London: Routledge.

————. 2004. *Who Owns the Past? The Politics of Time in a "Model" Bulgarian Village.* Oxford: Berghahn Books.

Kanev, Konstantin. 1975. *Minaloto na celo Momchilovtsi, Smolyansko: Prinos kum istoriyata na srednite Rodopi* (The past of the village of Momchilovtsi, Smolyan region: A contribution to the history of the middle Rhodope). Sofia: Izdatelstvo na otechestveniya front.

Karaivanov, Kostadin. 1990. *Gorite v tsentralni Rodopi* (The forests of the central Rhodope). Plovdiv: Izdatelstvo Hristo G. Danov.

Karov, Dimitur Damyanov. 1993. *Bilo li e. . .* (The way it was . . .). Sofia: ET Magi-Margarita Karova.

Keane, John, ed. 1998. *Civil Society and the State: New European Perspectives.* London: University of Westminster Press.

Kemf, Elizabeth. 1997. *WWF's Global Conservation Programme 1997/98.* Gland: WWF-International.

Kideckel, David A. 1993a. Once Again, the Land: Decollectivization and Social Conflict in Rural Romania. In *The Curtain Rises: Rethinking Culture, Ideology, and the State in Eastern Europe.* Hermine G. De Soto and David G. Anderson, eds. Pp. 62–77. Atlantic Highlands NJ: Humanities Press.

————. 1993b. *The Solitude of Collectivism: Romanian Villagers to the Revolution and Beyond.* Ithaca NY: Cornell University Press.

————. 1995a. Two Incidents on the Plains in Southern Transylvania: Pitfalls of Privatization in a Romanian Community. In *East European Communities: The Struggle for Balance in Turbulent Times.* David A. Kideckel, ed. Pp. 47–63. Boulder CO: Westview Press.

Kideckel, David A., ed. 1995b. *East European Communities: The Struggle for Balance in Turbulent Times*. Boulder CO: Westview Press.

Kipnis, Andrew B. 1997. *Producing Guanxi: Sentiment, Self, and Subculture in a North China Village*. Durham NC: Duke University Press.

Kligman, Gail, and Katherine Verdery. 1999. Reflections on the "Revolutions" of 1989 and After. *East European Politics and Societies* 13(2):303–312.

Knight, C. Gregory, and Marieta P. Staneva. 1996. The Water Resources of Bulgaria: An Overview. *GeoJournal* 40(4):347–362.

Kodjabashev, Alexander. 1995. Bulgaria. In *Status of Public Participation Practices in Environmental Decisionmaking in Central and Eastern Europe: Case Studies of Albanian, Bulgaria, the Czech Republic, Croatia, Estonia, Hungary, Latvia, Lithuania, FYR Macedonia, Poland, Romania, the Slovak Republic, and Slovenia*. Pp. 39–46. Budapest: Regional Environmental Center for Central and Eastern Europe.

Konstantinov, Yulian. 1992. An Account of Pomak Conversions in Bulgaria (1912–1990). In *Minderheitenfragen in Südosteuropa*. Gerhard Seewann, ed. Pp. 343–359. München: Südost-Institute, R. Oldenbourg Verlag.

———. 1993. Minority Problems of Self-Definition: Conventional and Minority Representations. In *Ethnicity and Politics in Bulgaria and Israel*. Jon Anson, Elka Todorova, Gideon Kressel, and Nikolai Genov, eds. Pp. 66–80. Aldershot: Avebury.

———. 1995. The Dragon of Kovachitsa—Local Perceptions of Radioactive Pollution near the Kozlodui Nuclear-Power Station (Bulgaria). *Human Ecology* 23(1):99–110.

———. 1997. Strategies for Sustaining a Vulnerable Identity: The Case of the Bulgarian Pomaks. In *Muslim Identity and the Balkan State*. Hugh Poulton and Suha Taij-Farouki, eds. Pp. 33–54. London: Hurst and Co. in association with the Islamic Council.

Korten, David C. 1990. *Getting to the 21st Century: Voluntary Action and the Global Agenda*. Hartford CT: Kumarian Press.

Kostova, Dobrinka. 2000. Bulgaria: Economic Elite Change during the 1990s. In *Elites after State Socialism: Theories and Analysis*. John Higley and György Lengyel, eds. Pp. 199–207. Lanham MD: Rowman and Littlefield.

Kottak, Conrad P., and Alberto G. G. Costa. 1993. Ecological Awareness, Environmentalist Action, and International Conservation Strategy. *Human Organization* 52(4):335–343.

Koulov, Boian. 1998. Post-Communist Change in Ecopolitics and Environmental Management: A Case Study of Bulgaria's Burgas Region. In *Bulgaria in Transition: Environmental Consequences of Political and Economic Transformation*. Krassimira Paskaleva, Philip Shapira, John Pickles, and Boian Koulov, eds. Pp. 201–228. Aldershot: Ashgate.

Krustanova, Kipriyana. 1986. *Traditsii na trudova vzaimopomoshto v bul-garskoto selo* (Traditions of labor mutual assistance in the Bulgarian village). Sofia: Izdatelstvo na Bulgarskata Akademiya na Naukite.

Kuipers, Sophie Emma. 1997. Trade in Medicinal Plants. In *Medicinal Plants for Forest Conservation and Health Care*. Non-Wood Forest Products Series 11. Gerard Bodeker, K. K. S. Bhat, Jeffrey Burley, and Paul Vantomme, eds. Pp. 45–59. Rome: Food and Agriculture Organization of the United Nations.

Kuzmanov, Georgi T. 1967. *Kooperativnoto dvizhenie v Smolyanski okrug* (The cooperative movement in the Smolyan region). Sofia: DPK Dimitur Blagoev.

Kyutchukov, Stephan. 1995. Bulgaria: Country Report. In *Selected Legislative Texts and Commentaries on Central and East European Not-for-Profit Law*. Douglas B. Rutzen, ed. Pp. 1–12. Sofia: International Center for Not-for-Profit Law, European Foundation Center, and Union of Bulgarian Foundations.

Lampe, John R. 1986. *The Bulgarian Economy in the Twentieth Century*. New York: St. Martin's Press.

Lampe, John R., and Marvin R. Jackson. 1982. *Balkan Economic History, 1550–1950: From Imperial Borderlands to Developing Nations*. Bloomington: Indiana University Press.

Lampland, Martha. 1995. *The Object of Labor: Commodification in Socialist Hungary*. Chicago: University of Chicago Press.

———. 2002. The Advantages of Being Collectivized: Cooperative Farm Managers in the Postsocialist Economy. In *Postsocialism: Ideals, Ideologies, and Practices in Eurasia*. C. M. Hann, ed. Pp. 31–56. London: Routledge.

Lange, Dagmar, and Magdalena Mladenova. 1997. Bulgarian Model for Regulating the Trade in Plant Material for Medicinal and Other Purposes. In *Medicinal Plants for Forest Conservation and Health Care*. Non-Wood Forest Products Series 11. Gerard Bodeker, K. K. S. Bhat, Jeffrey Burley, and Paul Vantomme, eds. Pp. 135–146. Rome: Food and Agriculture Organization of the United Nations.

Lange, Dagmar, and Uwe Shippmann. 1997. *Trade Survey of Medicinal Plants in Germany: A Contribution to International Plant Species Conservation*. Bonn: Bundesamt für Naturschutz.

Ledeneva, Alena V. 1998. *Russia's Economy of Favours: Blat, Networking and Informal Exchange*. Cambridge: Cambridge University Press.

Leonard, Pamela, and Deema Kaneff, eds. 2002. *Post-Socialist Peasant? Rural and Urban Constructions of Identity in Eastern Europe, East Asia, and the Former Soviet Union*. Basingstoke: Palgrave.

Lewis, David. 1999. Revealing, Widening, Deepening? A Review of the

Existing and Potential Contributions of Anthropological Approaches to "Third-Sector" Research. *Human Organization* 58(1):73–81.

Little, Peter D. 1994. The Link between Local Participation and Improved Conservation: A Review of Issues and Experiences. In *Natural Connections: Perspectives in Community-based Conservation*. David Western, R. Michael Wright, and Shirley C. Strum, eds. Pp. 347–372. Washington DC: Island Press.

Little, Peter D., and Michael M Horowitz. 1987. Introduction: Perspectives on Land, Ecology, and Development. In *Lands at Risk in the Third World: Local-Level Perspectives*. Peter D. Little and Michael M Horowitz, with A. Endre Nyerges, eds. Pp. 1–16. Boulder CO: Westview Press.

Livernash, Robert. 1992. The Growing Influence of NGOs in the Developing World. *Environment* 34(5):12–20, 41–43.

Lockwood, William G. 1973. The Peasant-Worker in Yugoslavia. *Studies in European Society* 1(1):91–110.

Machlis, Gary E., and David L. Tichnell. 1985. *The State of the World's Parks: An International Assessment for Research Management, Policy, and Research*. Boulder CO: Westview Press.

Malinowski, Bronislaw. 1922. *Argonauts of the Western Pacific*. London: Routledge.

Mandel, Ruth. 2002. Seeding Civil Society. In *Postsocialism: Ideals, Ideologies, and Practices in Eurasia*. C. M. Hann, ed. Pp. 279–296. London: Routledge.

Marcus, George E. 1995. Ethnography in/of the World System: The Emergence of Multi-Sited Ethnography. *Annual Review of Anthropology* 24:95–117.

Marcus, Richard R. 2001. Seeing the Forest for the Trees: Integrated Conservation and Development Projects and Local Perceptions of Conservation in Madagascar. *Human Ecology* 29(4):381–397

Markowitz, Lisa. 2001. Finding the Field: Notes on the Ethnography of NGOs. *Human Organization* 60(1):40–46.

Massam, Bryan H., and Robert Earl-Goulet. 1997. Environmental Nongovernmental Organizations in Central and Eastern Europe. *International Environmental Affairs* 9(2):127–147.

McCabe, J. Terrence, Scott Perkin, and Claire Schofield. 1992. Can Conservation and Development Be Coupled among Pastoral People? An Examination of the Maasai of the Ngorongoro Conservation Area, Tanzania. *Human Organization* 51(4):353–366.

McCay, Bonnie J. 1992. Everyone's Concern, Whose Responsibility? The Problem of the Commons. In *Understanding Economic Process*. Monographs in Economic Anthropology 10. Sutti Ortiz and Susan Lees, eds. Pp. 189–210. Lanham MD: University Press of America.

McCay, Bonnie J., and James M. Acheson, eds. 1987. *The Question of the Commons: The Culture and Ecology of Communal Resources.* Tucson: University of Arizona Press.

McGrew, William W. 1985. *Land and Revolution in Modern Greece, 1800–1881: The Transition in the Tenure and Exploitation of Land from Ottoman Rule to Independence.* Kent OH: Kent State University Press.

McIntyre, Robert J. 1980. The Bulgarian Anomaly: Demographic Transition and Current Fertility. *Southeastern Europe* 7(2):147–170.

———. 1988. *Bulgaria: Politics, Economics and Society.* London: Pinter.

McNeely, Jeffrey A. 1993. Fourth World Congress on National Parks and Protected Areas, Held in Caracas, Venezuela, during 10–21 February 1992. *Environmental Conservation* 20(1):89.

———. 1994. Protected Areas for the 21st Century: Working to Provide Benefits to Society. *Biodiversity and Conservation* 3:390–405.

———. 1999. *Mobilizing Broader Support for Asia's Biodiversity: How Civil Society Can Contribute to Protected Area Management.* Manila: Asian Development Bank.

Mehta, J. N., and S. R. Kellert. 1998. Local Attitudes toward Community-Based Conservation Policy and Programmes in Nepal: A Case Study in the Makalu-Barun Conservation Area. *Environmental Conservation* 25(4):320–333.

Meine, Curt. 1994. *Conserving Biological Diversity in Bulgaria: The National Biological Diversity Conservation Strategy.* Washington DC: Biodiversity Support Program.

Meine, Curt, ed. 1998. *Bulgaria's Biological Diversity: Conservation Status and Needs Assessment,* vols. 1 and 2. Washington DC: Biodiversity Support Program.

Meurs, Mieke, and Darren Spreeuw. 1997. Evolution of Agrarian Institutions in Bulgaria: Markets, Cooperatives, and Private Farming, 1991–94. In *The Bulgarian Economy: Lessons from Reform during Early Transition.* Derek C. Jones and Jeffrey Miller, eds. Pp. 275–297. Aldershot: Ashgate.

Meurs, Mieke, and Simeon Djankov. 1998. The Alchemy of Reform: Bulgarian Agriculture in the 1980s. In *Privatizing the Land: Rural Political Economy in Post-Communist Societies.* Iván Szelényi, ed. Pp. 43–61. London: Routledge.

Meyer, Carrie A. 1995. Opportunism and NGOs: Entrepreneurship and Green North–South Transfers. *World Development* 23(8):1277–1289.

Michev, Tanyu, and Petar Iankov. 1998. The Bulgarian Ornithofauna. In *Bulgaria's Biological Diversity: Conservation Status and Needs Assessment,* vols. 1 and 2. Curt Meine, ed. Pp. 411–438. Washington DC: Biodiversity Support Program.

Mihova, Boriana. 1998. Summary Report of the Bulgarian Conservation

Non-Governmental Organizations. In *Bulgaria's Biological Diversity: Conservation Status and Needs Assessment*, vols. 1 and 2. Curt Meine, ed. Pp. 703–717. Washington DC: Biodiversity Support Program.

Miller, Jeffrey B. 1997. Price Index Gap in Bulgaria. In *The Bulgarian Economy: Lessons from Reform during Early Transition*. Derek C. Jones and Jeffrey Miller, eds. Pp. 63–80. Aldershot: Ashgate.

Miller, Kenton R. 1984. The Natural Protected Areas of the World. In *National Parks, Conservation and Development: The Role of Protected Areas in Sustaining Society*. Jeffrey A. McNeely and Kenton R. Miller, eds. Pp. 20–23. Washington DC: Smithsonian Institution Press.

Ministerstvo na Teritorialnoto Razvitie i Stroitelstvoto. 1994. *Planinskite rayoni v republika Bulgariya: Obzor, politika, namerenia* (The mountain regions of the Republic of Bulgaria: Survey, policy, findings). Sofia: Ministerstvo na Teritorialnoto Razvitie i Stroitelstvoto.

Ministry of Environment. 1995. *National Action Plan for the Conservation of the Most Important Wetlands in Bulgaria*. Sofia: Ministry of Environment.

Ministry of Environment and Waters, Republic of Bulgaria. 2000. *The National Biodiversity Conservation Plan*. Sofia: MOEW.

Mitsuda, Hisayoshi, and Konstantin Pashev. 1995. Environmentalism as Ends or Means? The Rise and Political Crisis of the Environmental Movement in Bulgaria. *Capitalism Nature Society* 6(1):87–111.

Monga, Celestin. 1996. *The Anthropology of Anger: Civil Society and Democracy in Africa*. Boulder CO: Lynne Rienner Publishers.

Monov, Tsvyatko. 1983. Ikonomicheski i sotsialni izmeneniya v Rodopskiya kray (1944–1977) (Economic and social change in the Rhodope region [1944–1977]). In *Rodopski Sbornik*, vol. 5. Hr. Hristov, P. Petrov, and Str. Dimitrov, eds. Pp. 5–40. Sofia: Izdatelstvo na Bulgarskata Akademiya na Naukite.

———. 1987. Selskoto stopanstvo i agrarnata struktura v rodopite prez epohata na kapitalizma (1918–1944) (The rural economy and the agrarian structure of the Rhodope during the epoch of capitalism [1918–1944]). In *Rodopski Sbornik*, vol. 6. Hr. Hristov, P. Petrov, and Str. Dimitrov, eds. Pp. 54–82. Sofia: Izdatelstvo na Bulgarskata Akademiya na Naukite.

Murphree, Marshall W. 1994. The Role of Institutions in Community-Based Conservation. In *Natural Connections: Perspectives in Community-based Conservation*. David Western, R. Michael Wright, and Shirley C. Strum, eds. Pp. 403–427. Washington DC: Island Press.

Nagengast, Carole. 1991. *Reluctant Socialists, Rural Entrepreneurs: Class, Culture, and the Polish State*. Boulder CO: Westview Press.

National Statistical Institute, Republic of Bulgaria. 1994. *Rezultati ot prebroyavaneto na nacelenieto. tom III. Broi na nacelenieto po oblasti, obshtini*

i naceleni mesta (okonchaltelni danni) (Results of the census of population, vol 3: Population by districts, municipalities, and populated
   places [final data]). Sofia: Statistichesko izdatelstvoi pechatnitsa.

———. 1996a. *Statistical Yearbook 1996.* Sofia: Statistichesko izdatelstvo i
   pechatnitsa.

———. 1996b. *Sotsialno-demografski harakteristiki na nacelenieto i zhilishten
   fond, region Smolyan* (Socio-demographic characteristics of the population and housing fund, Smolyan region). Sofia: Statistichesko izdatelstvo i pechatnitsa.

———. 1997. *Biudjeti na domakinstvata v Republika Bulgariya* (Household
   budgets in Bulgaria). Sofia: Statistichesko izdatelstvo i pechatnitsa.

———. 1998. *Bulgaria '98.* Electronic document, http://w3.nis.bg/cgi-
   bin/sel.cgi/publ/chapt/ ?FPUBL=6, accessed June 1999.

National Trust EcoFund, Republic of Bulgaria. 1998. General Information. Unpublished mimeographed document, Sofia. Obtained from
   the office of the National Trust EcoFund.

Ndegwa, Stephen. 1996. *The Two Faces of Civil Society: NGOs and Politics
   in Africa.* West Hartford CT: Kumarian Press.

Netting, Robert McC. 1976. What Alpine Peasants Have in Common:
   Observations on Communal Tenure in a Swiss Village. *Human Ecology*
   4(2):135–146.

———. 1981. *Balancing on an Alp: Ecological Change and Continuity in a
   Swiss Mountain Community.* Cambridge: Cambridge University Press.

Neumann, Roderick P. 1998. *Imposing Wilderness: Struggles over Livelihood
   and Nature Preservation in Africa.* Berkeley: University of California
   Press.

Nikolov, Stephan E. 1992. The Emerging Nonprofit Sector in Bulgaria:
   Its Historical Dimensions. In *The Nonprofit Sector in the Global Com-
   munity: Voices from Many Nations.* Kathleen D. McCarthy, Virginia A.
   Hodgkinson, Russy D. Sumariwalla, and associates, eds. Pp. 333–348.
   San Francisco: Jossey-Bass Publishers.

———. 1999. *Blagotvoryashtiyat sektor* (The benevolent sector). Sofia:
   P. I. K. Litera.

Nikolova, Mariana. 1998. Technological Hazards in Bulgaria. In *Bulgaria
   in Transition: Environmental Consequences of Political and Economic Transformation.* Krassimira Paskaleva, Philip Shapira, John Pickles, and
   Boian Koulov, eds. Pp. 39–56. Aldershot: Ashgate.

Nygren, A. 2000. Environmental Narratives on Protection and Production: Nature-based Conflicts in Rio San Juan, Nicaragua. *Development
   and Change* 31(4):807–830.

Oli, M. K., I. R. Taylor, and M. E. Rogers. 1994. Snow Leopard *Panthera
   uncia* Predation of Livestock: An Assessment of Local Perceptions in

the Annapurna Conservation Area, Nepal. *Biological Conservation* 68(1):63–68.

Open Media Research Institute (OMRI). 1997. Protests Continue in Sofia. *OMRI Daily Digest*, January 15.

Organizirane iznosa na lukoviti trevi i bilki (Organized export of medicinal plants and herbs). 1939. *Rodopska Iskra*, July 10 (51/52):2–3.

Orlove, Benjamin. 1986. Barter and Cash Sale on Lake Titicaca: A Test of Competing Approaches. *Current Anthropology* 27:85–106.

Orlove, Benjamin S., and Stephen B. Brush. 1996. Anthropology and the Conservation of Biodiversity. *Annual Review of Anthropology* 25:329–352.

Ostrom, Elinor. 1990. *Governing the Commons: The Evolution of Institutions for Collective Action*. Cambridge: Cambridge University Press.

Parman, Susan. 1998. Introduction: Europe in the Anthropological Imagination. In *Europe in the Anthropological Imagination: Exploring Cultures*. Susan Parman, ed. Pp. 1–16. Upper Saddle River NJ: Prentice Hall.

Peet, Richard, and Michael Watts. 1996. Liberation Ecology: Development, Sustainability, and Environment in an Age of Market Triumphalism. In *Liberation Ecologies: Environment, Development, Social Movements*. Richard Peet and Michael Watts, eds. Pp. 1–45. London: Routledge.

Peev, Dimitar, Tenyu Meshinev, Nikolaj Spassov, Jeko Spiridonov, Lyubomira Mileva, Petar Yankov, Lyubomir Profirov, Velitchko Velitchkov, Maria Karapetkova, and Lyubomir Andreev. 1995. *Bulgaria: Natural Heritage*. Sofia: Tilia.

Pelto, Pertti J., and Billie R. DeWalt. 1985. Methodology in Macro-Micro Studies. In *Micro and Macro Levels of Analysis in Anthropology*. Billie R. DeWalt and Pertti J. Pelto, eds. Pp. 187–201. Boulder CO: Westview Press.

Penchovska, Julietta. N.d. Analiz na anketa provedena sred nepravitelstveni organizatsii za opazvane na okolnata sreda v Bulgariya (Analysis of inquiry held among nongovernmental organizations for the protection of the environment in Bulgaria). Unpublished MS.

Penchovska, Julietta, Dragomir Petrov, and Antoaneta Kobakova. 1997. *Spravochnik na nepravitelstvenite organizatsii za opazvane na okolnata sreda* (Catalogue of environmental nongovernmental organizations in Bulgaria). Sofia: Regional Environmental Center for Central and Eastern Europe.

Penchovska, Julietta, Mariana Ivanova, and Antoaneta Kobakova. 1993. *Catalogue of Environmental Non-Governmental Organizations in Bulgaria*. Sofia: Regional Environmental Center for Central and Eastern Europe.

Perry, Julian. 1995. *The Mountains of Bulgaria: A Walker's Companion.* Leicester: Cordee.

Peters, Pauline E. 1994. *Dividing the Commons: Politics, Policy, and Culture in Botswana.* Charlottesville: University Press of Virginia.

Pickles, John, and the Bourgas Group. 1993. Environmental Politics, Democracy, and Economic Restructuring in Bulgaria. In *The New Political Geography of Eastern Europe.* John O'Loughlin and Herman van der Wusten, eds. Pp. 167–185. London: Belhaven.

Pickles, John, and Dimitrina Mikhova. 1998. The Political Economy of Environmental Data in the Bulgarian Transformation. In *Bulgaria in Transition: Environmental Consequences of Political and Economic Transformation.* Krassimira Paskaleva, Philip Shapira, John Pickles, and Boian Koulov, eds. Pp. 105–127. Aldershot: Ashgate.

Pickles, John, Petr Pavlínek, and Caedmon Staddon. 1998. Who Cares about the Environment? Popular Attitudes to Pollution and Environmental Reconstruction: Two Case Studies of Bulgaria and the Czech Republic. In *Bulgaria in Transition: Environmental Consequences of Political and Economic Transformation.* Krassimira Paskaleva, Philip Shapira, John Pickles, and Boian Koulov, eds. Pp. 229–256. Aldershot: Ashgate.

Pickvance, Katy. 1998. *Democracy and Environmental Movements in Eastern Europe: A Comparative Study of Hungary and Russia.* Boulder CO: Westview Press.

Pine, Frances, and Sue Bridger. 1998. Introduction: Transitions to Post-Socialism and Cultures of Survival. In *Surviving Post-Socialism: Local Strategies and Regional Responses.* Sue Bridger and Frances Pine, eds. Pp. 1–15. London: Routledge.

Price, Marie. 1994. Ecopolitics and Environmental Nongovernmental Organizations in Latin America. *Geographical Review* 84(1):42–58.

Primovski, Anastas. 1973. Bit i kultura na rodopskite Bulgari (Customs and culture of the Rhodope Bulgarians). *Sbornik za narodni umotvoreniya* 54.

Princen, Thomas, and Matthias Finger. 1994. *Environmental NGOs in World Politics: Linking the Local and the Global.* London: Routledge.

Pryor, Frederic L. 1977. *The Origins of the Economy.* New York: Academic Press.

Regional Environmental Center for Central and Eastern Europe (REC). 1993. *Introducing the REC.* Budapest: REC.

———. 1997. *Problems, Progress, and Possibilities: A Needs Assessment of Environmental NGOs in Central and Eastern Europe.* Szentendre: REC.

Rhoades, Robert E., and Stephen I. Thompson. 1975. Adaptive Strategies in Alpine Environments: Beyond Ecological Particularism. *American Ethnologist* 2:535–551.

Rizov, Kiril. 1987. *Razvitie na prirodozashtitnoto delo v Bulgariya* (The development of nature protection activities in Bulgaria). Sofia: Izdatelstvo na otechestveniya front.

Roha, Ronaleen R. 1996. How Bartering Saves Cash. *Kiplinger's Personal Finance Magazine* (February):103–107.

Róna-Tas, Ákos, and József Böröcz. 2000. Bulgaria, the Czech Republic, Hungary, and Poland: Presocialist and Socialist Legacies among Business Elites. In *Elites after State Socialism: Theories and Analysis.* John Higley and György Lengyel, eds. Pp. 209–227. Lanham MD: Rowman and Littlefield.

Roth, Klaus. 1990. Socialist Life-Cycle Rituals in Bulgaria. *Anthropology Today* 6(5):8–10.

Ruffin, M. Holt, ed. 1999. *Civil Society in Central Asia.* Seattle: University of Washington Press.

Sampson, Steven. 1993. Money without Culture, Culture without Money: Eastern Europe's Nouveaux Riches. *Anthropological Journal on European Cultures* 3(1):7–30.

———. 1995. All Is Possible, Nothing Is Certain: The Horizons of Transition in a Romanian Village. In *East European Communities: The Struggle for Balance in Turbulent Times.* David A. Kideckel, ed. Pp. 159–176. Boulder CO: Westview Press.

———. 1996. The Social Life of Projects: Importing Civil Society to Albania. In *Civil Society: Challenging Western Models.* C. M. Hann and Elizabeth Dunn, eds. Pp. 121–142. London: Routledge.

———. 2002. Beyond Transition: Rethinking Elite Configurations in the Balkans. In *Postsocialism: Ideals, Ideologies, and Practices in Eurasia.* C. M. Hann, ed. Pp. 297–316. London: Routledge.

Sanders, Irwin T. 1949. *Balkan Village.* Lexington: University of Kentucky Press.

Seabright, Paul, ed. 2000. *The Vanishing Rouble: Barter Networks and Non-Monetary Transactions in Post-Soviet Societies.* Cambridge: Cambridge University Press.

Sikor, Thomas. 2001. Agrarian Differentiation in Post-Socialist Societies: Evidence from Three Upland Villages in North-Western Vietnam. *Development and Change* 32(5):923–949.

Silverman, Carol. 1983. The Politics of Folklore in Bulgaria. *Anthropological Quarterly* 56(2):55–61.

———. 1984. Pomaks. In *Muslim Peoples: A World Ethnographic Survey.* 2nd edition. Richard V. Weekes, ed. Pp. 612–616. Westport CT: Greenwood Press.

———. 1986. Bulgarian Gypsies: Adaptation in a Socialist Context. *Nomadic Peoples* 21–22:51–62.

———. 2000. Researcher, Advocate, Friend: An American Fieldworker among Balkan Roma, 1980–1996. In *Fieldwork Dilemmas: Anthropologists in Postsocialist States.* Hermine G. De Soto and Nora Dudwick, eds. Pp. 195–217. Madison: University of Wisconsin Press.

Simons, Marlise. 1994. East Europe Still Choking on Air of the Past. *New York Times,* November 3:A1, A14.

Siuleymanov, Atem Smailov H. 1968. Istoricheski ocherk na selo Zaburdo—Smolyansko (A historical sketch of the village of Zaburdo, Smolyan region). Unpublished thesis, Sofiyski Universitat "Kliment Ohridski."

Słomczyński, Kazimierz M., and Goldie Shabad. 1997. Systemic Transformation and the Salience of Class Structure in East Central Europe. *East European Politics and Societies* 11(1): 155–189.

Smart, Alan. 1993. Gifts, Bribes, and Guanxi: A Reconsideration of Bourdieu's Social Capital. *Cultural Anthropology* 8(3):388–408.

Smollett, Eleanor. 1980. Implications of the Multicommunity Production Cooperative (Agro-Industrial Complex) for Rural Life in Bulgaria or the Demise of the Kara Stoyanka. *Bulgarian Journal of Sociology* 3:42–56.

———. 1989. The Economy of Jars: Kindred Relationships in Bulgaria—An Exploration. *Ethnologia Europaea* 19:125–140.

Smolyan Territorial Statistical Bureau, National Statistical Institute. 1995. *Statisticheski sbornik, Smolyan '94* (Statistical handbook, Smolyan '94). Smolyan: Statistichesko izdatelstvo i pechatnitsa.

———. 1997. *Statisticheski sbornik, Smolyan '96* (Statistical handbook, Smolyan '96). Smolyan: Statistichesko izdatelstvo i pechatnitsa.

Snavely, Keith. 1996. The Welfare State and the Emerging Non-Profit Sector in Bulgaria. *Europe-Asia Studies* 48(4):647–662.

Snavely, Keith, and Uday Desai. 1995. Bulgaria's Nonprofit Sector: The Search for Form, Purpose, and Legitimacy. *Voluntas* 6:23–38.

Soil Characteristics and Soil Map of the Territory of the village of Zaburdo, Smolyan Region. 1973. Unpublished document from the Cartographic Archive of the N. Poushkarov Institute of Soil Science and Agroecology.

Staddon, Caedmon Bertrand. 1996. Democratisation, Environmental Management, and the Production of New Political Geographies in Bulgaria: A Case Study of the 1994–95 Sofia Water Crisis. Ph.D. dissertation, University of Kentucky.

Staddon, Caedmon, and Barbara Cellarius. 2002. Paradoxes of Conservation and Development in Postsocialist Bulgaria: Recent Controversies. *European Environment* 12(2):105–116.

Staver, Charles, Robert Simeone, and Anthony Stocks. 1994. Land Resource Management and Forest Conservation in Central Amazonian

Peru: Regional, Community, and Farm-Level Approaches among Native Peoples. *Mountain Research and Development* 14(2):147–157.

Stevens, Stanley F. 1993. *Claiming the High Ground: Sherpas, Subsistence, and Environmental Change in the Highest Himalaya*. Berkeley: University of California Press.

Stevens, Stanley F., and Terry De Lacy, eds. 1997. *Conservation through Cultural Survival: Indigenous People and Protected Areas*. Washington DC: Island Press.

Stillman, Edmund O. 1958. The Collectivization of Bulgarian Agriculture. In *The Collectivization of Agriculture in Eastern Europe*. Enno E. Kraene, Philip E. Mosely, Edmund O. Stillman, Ernst Koenig, Nicolas Spulber, Jozo Tomasevich, and Irwin T. Sanders, eds. Pp. 67–102. Lexington: University of Kentucky Press.

Stoyanov, V. 1968. *Istoriya na gorskoto stopanstvo na Bulgariya, Chast 1* (History of forestry in Bulgaria, vol. 1). Sofia: Izdatelstvo na Bulgarskata Akademiya na Naukite.

Strong, Ann Louise, Thomas A. Reiner, and Janusz Szyrmer. 1996. *Transitions in Land and Housing: Bulgaria, The Czech Republic, and Poland*. New York: St. Martin's Press.

Suiuz za zashtita na prirodata. 1995. Zarazhdane i razvitie na prirodozashtitnoto dvizhenie v Bulgariya i suiuza za zashtita na prirodata (Origin and development of the nature protection movement in Bulgaria and the Union for the Protection of Nature). *Biuletin* 1 (July 7, 1995):1–3.

Szelényi, Iván, and Szonja Szelényi. 1995. Circulation or Reproduction of Elites during Post-Communist Transformations in Eastern Europe: Introduction. *Theory and Society* 24 (October):615–638.

Szelényi, Szonja, in collaboration with Karen Aschaffenburg, Mariko Lin Chang, and Winifred Poster. 1998. *Equality by Design: The Grand Experiment in Destratification in Socialist Hungary*. Palo Alto CA: Stanford University Press.

Tendler, Judith. 1982. Turning Private Voluntary Organizations into Development Agencies: Questions for Evaluation. AID Program Evaluation Discussion Paper 12. Washington DC: U.S. Agency for International Development.

Thomas-Slayer, Barbara P. 1992. Implementing Effective Local Management of Natural Resources: New Roles for NGOs in Africa. *Human Organization* 51(2):136–143.

Tismaneanu, Vladimir, ed. 1995. *Political Culture and Civil Society in Russia and the New States of Eurasia*. Armonk NY: Sharpe.

Todorova, Maria. 1997. Identity (Trans)Formation among Pomaks in Bulgaria. In *Beyond Borders: Remaking Cultural Identities in the New East and Central Europe*. Laszlo Kurti and Juliet Langman, eds. Pp. 63–82. Boulder CO: Westview Press.

Tong, Yanqi. 1995. Mass Alienation under State Socialism and After. *Communist and Post-Communist Studies* 28(2):215–237.

Tsentralno Statistichesko Upravlenie, Narodna Republika Bulgariya. 1988. *Demografska i sotsialno-ikonomicheska harakteristika na nacelenieto, Plovdivska oblast. tom I.* (Demographic and socio-economic characteristics of the population, Plovdiv district, vol. 1). Sofia: Tsentralno Statistichesko Upravlenie.

Turnbull, Colin M. 1972. *The Mountain People*. New York: Simon and Schuster.

Tzanov, Vassil, and Daniel Vaughan-Whitehead. 1997. Macroeconomic Effects of Restrictive Wage Policy in Bulgaria: Empirical Evidence for 1991–1995. In *The Bulgarian Economy: Lessons from Reform during Early Transition*. Derek C. Jones and Jeffrey Miller, eds. Pp. 99–125. Aldershot: Ashgate.

UNESCO World Heritage Centre. 1999. Properties Included in the World Heritage List. Electronic document, http://www.unesco.org/whc/archive/list99-eng.pdf, accessed March 25, 2004.

United Nations Development Programme (UNDP). 2001. *Citizen Participation in Governance: From Individuals to Citizens*. National Human Development Report, Bulgaria, 2001. Sofia: UNDP.

United States General Accounting Office (USGAO). 1994. *Environmental Issues in Central and Eastern Europe: U.S. Efforts to Help Resolve Institutional and Financial Problems*. Washington DC: Government Printing Office.

Uphoff, Norman. 1996. Why NGOs Are Not a Third Sector: A Sectoral Analysis with Some Thoughts on Accountability, Sustainability, and Evaluation. In *Beyond the Magic Bullet: NGO Performance and Accountability in the Post–Cold War World*. Michael Edwards and David Hulme, eds. Pp. 23–39. West Hartford CT: Kumarian Press.

Vakarelski, Hristo. 1969. Pominutsi na Bulgarite mohamedani i hristiyani v srednite rodopi (The occupations of the Bulgarian Muslims and Christians in the middle Rhodope). *Izvestiya na ethnografskiya institut i muzey* 12:39–68.

Vakil, Anna C. 1997. Confronting the Classification Problem: Toward a Taxonomy of NGOs. *World Development* 25(12):2057–2070.

Vandergeest, P. 1996. Property Rights in Protected Areas: Obstacles to Community Involvement as a Solution in Thailand. *Environmental Conservation* 23(3):259–268.

Vari, Anna, and Pal Tamas, eds. 1993. *Environment and Democratic Transitions: Policy and Politics in Central and Eastern Europe*. Dordrecht: Kluwer Academic Publishers.

Vayda, Andrew P., and Bradley B. Walters. 1999. Against Political Ecology. *Human Ecology* 27(1):167–180.

Velev, Stefan B. 1996. Is Bulgaria Becoming Warmer and Drier? *GeoJournal* 40(4):363–370.

Verdery, Katherine. 1991. Theorizing Socialism: A Prologue to the "Transition." *American Ethnologist* 18(3):419–439.

———. 1993. Ethnic Relations, Economies of Shortage, and the Transition in Eastern Europe. In *Socialism: Ideals, Ideologies, and Local Practice.* C. M. Hann, ed. Pp. 172–186. London: Routledge.

———. 1994. The Elasticity of Land: Problems of Property Restitution in Transylvania. *Slavic Review* 53(4):1071–1109.

———. 1996. *What Was Socialism and What Comes Next?* Princeton NJ: Princeton University Press.

———. 1998. Property and Power in Transylvania's Decollectivization. In *Property Relations: Renewing the Anthropological Tradition.* C. M. Hann, ed. Pp. 160–180. Cambridge: Cambridge University Press.

———. 1999. Fuzzy Property: Rights, Power, and Identity in Transylvania's Decollectivization. In *Uncertain Transition: Ethnographies of Change in the Postsocialist World.* Michael Burawoy and Katherine Verdery, eds. Pp. 53–82. Lanham MD: Rowman and Littlefield.

Vivian, Jessica. 1994. NGOs and Sustainable Development in Zimbabwe: No Magic Bullets. *Development and Change* 25:167–193.

Wedel, Janine R. 1998a. *Collision and Collusion: The Strange Case of Western Aid to Eastern Europe 1989–1998.* New York: St. Martin's Press.

———. 1998b. Informal Relations and Institutional Change: How Eastern European Cliques and States Mutually Respond. *Anthropology of East Europe Review* 16(1):4–14.

Weiner, Douglas R. 1999. *A Little Corner of Freedom: Russian Nature Protection from Stalin to Gorbachev.* Berkeley: University of California Press

Wells, Michael P., and Katrina E. Brandon. 1992. *People and Parks: Linking Protected Area Management with Local Communities.* Washington DC: World Bank.

Wells, Michael P., and Margaret D. Williams. 1998. Russia's Protected Areas in Transition: The Impacts of Perestroika, Economic Reform, and the Move towards Democracy. *Ambio* 27(3):198–206.

Werner, Cynthia. 2000. Gifts, Bribes, and Development in Post-Soviet Kazakstan. *Human Organization* 59(1):11–22.

West, Patrick C., and Steven R. Brechin, eds. 1991. *Resident Peoples and National Parks: Social Dilemmas and Strategies in International Conservation.* Tucson: University of Arizona Press.

Western, David. 1993. Ecosystem Conservation and Rural Development: The Amboseli Case. Paper prepared for the Liz Claiborne–Art

Ortenberg Foundation Community-based Conservation Workshop. Airlie, Virginia, 18–22 October 1993.

Western, David, R. Michael Wright, and Shirley C. Strum, eds. 1994. *Natural Connections: Perspectives in Community-based Conservation.* Washington DC: Island Press.

Whitaker, Roger. 1979. Continuity and Change in Two Bulgarian Communities: A Sociological Profile. *Slavic Review* 38(2):259–271.

Wilson, Edward O. 1992. *The Diversity of Life.* Cambridge MA: Harvard University Press.

Wolf, Eric. 1972. Ownership and Political Ecology. *Anthropological Quarterly* 45(3):201–205.

Woodruff, David. 1999. Barter of the Bankrupt: The Politics of Demonetization in Russia's Federal State. In *Uncertain Transition: Ethnographies of Change in the Postsocialist World.* Michael Burawoy and Katherine Verdery, eds. Pp. 83–124. Lanham MD: Rowman and Littlefield.

World Bank. 1994. *Bulgarian Environmental Strategy Study: Update and Follow-up.* Washington DC: World Bank.

Wyzan, Michael L. 1997. Economic Transformation and Regional Inequality in Bulgaria: In Search of a Meaningful Unit of Analysis. In *The Bulgarian Economy: Lessons from Reform during Early Transition.* Derek C. Jones and Jeffrey Miller, eds. Pp. 21–49. Aldershot: Ashgate.

————. 1998. Bulgarian Economic Policy and Performance, 1991–1997. In *Bulgaria in Transition: Politics, Economics, Society, and Culture after Communism.* John D. Bell, ed. Pp. 93–122. Boulder CO: Westview Press.

Yankov, Peter. 1998. Report of the Bulgarian Bird Protection Society. In *Bulgaria's Biological Diversity: Conservation Status and Needs Assessment,* vols. 1 and 2. Curt Meine, ed. Pp. 727–740. Washington DC: Biodiversity Support Program.

Yoder, Jennifer A. 1999. *From East Germans to Germans? The New Postcommunist Elites.* Durham NC: Duke University Press.

Zbierski-Salameh, Slawomira. 1999. Polish Peasants in the "Valley of Transition": Responses to Postsocialist Reforms. In *Uncertain Transition: Ethnographies of Change in the Postsocialist World.* Michael Burawoy and Katherine Verdery, eds. Pp. 189–222. Lanham MD: Rowman and Littlefield.

Zerner, Charles, ed. 2000. *People, Plants, and Justice: The Politics of Nature Conservation.* New York: Columbia University Press.

Zhelyazkova, Antonina. 1998. Bulgaria's Muslim Minorities. In *Bulgaria in Transition: Politics, Economics, Society, and Culture after Communism.* John D. Bell, ed. Pp. 165–187. Boulder CO: Westview Press.

Index

agrarian party. *See* Bulgarian Agricultural National Union

agricultural chemicals, 95, 99, 114, 120, 121, 148

agricultural land: lack of a market for, 148; ownership of, 73, 87–88, 110–111, 113; restitution of, 109–111, 147–149

agricultural technology: postsocialist, 114, 116, 120, 134; presocialist, 88; socialist era, 99

agriculture: collectivization of, 97–98; constraints on, 148, 191; crop varieties, 89, 98 118–119; decision making in, 122; decollectivization of, 14; environmental conditions for, 105–109; labor organization in, 122–126; manure use in, 88–89, 92, 114, 115, 120, 136; mutual labor exchange in, 123–124; personal production (socialist era), 100–101; postsocialist, 112–126; presocialist, 88–90; product sales and marketing, 117–118; seasonal scheduling of activities, 175–178; socialist era, 94–95, 97–101; wage labor in, 124–125. *See also* animal husbandry

agro-industrial complexes, 94–95

akord system, 100, 101, 103

Albania, 62–63, 215

animal husbandry: decision making in, 126–127; herding practices, 91, 98, 129–134, 285 n14, 288 nn16, 17; livestock feed, 128, 129, 287–288 n15; mutual labor exchange in, 131–132; postsocialist, 126–136; presocialist, 90–92; product sales, 134–135; wage labor in, 132

anthropology: research on Bulgaria, 12–13, 279 n7; research on Central and Eastern Europe, 12–15, 279 n5; research on nature conservation, 8, 265–266; research on NGOs, 11–12

Arda River, 54, 268

assembly workshops. *See* rural industries

Bachkovo, 77

Balkan Mountains, 240, 282 n7

Bansko, 218

barter: during economic crisis, 169–170; general discussion of, 160–163; goods obtained through, 167; organization of, 167–169; in Zaburdo and similar villages, 159, 163–170, 289–290 n6

Bat Protection Society, 255

Batak, 235–238

biodiversity conservation, 5–9; challenges to, in Central and Eastern

cash-based exchange, 163, 171–174
caving clubs. *See* Chepelare: caving
club in; Trigrad caving club
Center for the Sustainable Development of the Mountains–Smolyan,
217, 221
Central Balkans National Park, 36, 38,
39, 40, 231
Chepelare, 72, 78; caving club in, 259,
294 n8; climate, 106–108, 286 n1;
forestry cooperatives in, 93; hunting and fishing society in, 146;
NGO location in, 65, 216, 225; ties
of Zaburdo assembly workshop
to, 103
Chervenata Stena Reserve, 70, 77
Citizens Initiative Committee for
the Protection of Rila Waters, 218,
221
Civil Committee for the Ecological
Defense of Ruse, 29
civil society: assumptions about the
role of NGOs in, 42, 43, 48, 50, 57,
248–249, 256; lessons from Bulgarian environmental NGOs, 270–271;
research on, 14–15
Colorado potato beetles, 121
Committee for Nature Protection,
31–32
Committee on Forests, 218
community property, 88
connections. *See* social networks
conservation. *See* biodiversity conservation; nature conservation
Consomol, 34
Constitution of Bulgaria, 15, 281 n4;
on environmental protection, 36,
46; on property, parks and reserves, 281 n4; on public participation, 51; on religious freedom,
284 n11
contract farming. *See* akord system
Convention for the Protection of the
World Cultural and Natural Heritage, 32, 39
cooperative, commercial in Zaburdo,
80, 96, 102, 146, 199, 291 n5

cooperative farm, 83, 97–99; liquidation of, 111–112
cooperatives, 285 n15; agricultural,
286 n5, 287–288 n15; forest, 93–94
corruption, 63–64, 171
Council for Mutual Economic Assistance, 16, 94
Council for the Protection of the
Countryside. *See* Union for the
Protection of Nature
Creed, Gerald, 13, 280 n7
crisis, economic and political, 17–20,
150; barter during, 169–170; impact on environmental NGOs,
61–66

debt-for-the-environment swap, 282–
283 n12
decollectivization. *See* agriculture:
decollectivization of; cooperative
farm: liquidation of
Democracy Network Program, 57–58,
224, 240
Devil's Throat Cave, 250, 253, 255, 258,
260
dowries, 155, 285 n13
Draginovo, 217
Dupkata Reserve, 76–77
Dutch Foundation for Education and
Training, 58–59

Ecoglasnost, 29, 41, 45, 46, 294 n9
Ecoglasnost-Burgas, 43
Ecoglasnost Independent Union, 49,
221
Ecoglasnost-National Movement, 46,
222
Ecoglasnost-Varna, 39
economic crisis. *See* crisis: economic
and political
economic diversification, 190–191
economy, importance of rural–urban
connections, 170–171; postsocialist, 150–192, 266–267, 291 n18;
presocialist, 85–94; socialist era,
94–95
economy of favors, 170–171

www.ingramcontent.com/pod-product-compliance
Lightning Source LLC
Chambersburg PA
CBHW050225270326
41914CB00003BA/568